TREATISE

ON

ALGEBRA.

BY ELIAS LOOMIS, A.M.,

PROFESSOR OF MATHEMATICS AND NATURAL PHILOSOPHY IN THE UNIVERSITY OF THE CITY OF NEW YORK, MEMBER OF THE AMERICAN PHILOSOPHICAL SOCIETY, OF THE AMERICAN ACADEMY OF ARTS AND SCIENCES, AND AUTHOR OF "A COURSE OF MATHEMATICS."

NEW YORK:
HARPER & BROTHERS, PUBLISHERS,
329 & 331 PEARL STREET
FRANKLIN SQUARE.
1852.

TO

THE COUNCIL

OF

THE UNIVERSITY OF THE CITY OF NEW YORK,

THIS TREATISE

Is respectfully Inscribed,

BY

THEIR OBEDIENT SERVANT,

THE AUTHOR.

PREFACE.

The first edition of my Algebra was received with unexpected favor. Almost immediately after its publication, it was adopted as a text-book in half a dozen colleges, besides numerous academies and schools; and the most flattering testimonials were received from every part of the country. I have thus been stimulated to increased exertions to render it less unworthy of public favor. Every line of it has been subjected to a thorough revision. The work has been read by two successive classes in the University, and wherever improvement seemed practicable, alterations have been freely made. I have also availed myself of the suggestions of several professors in other colleges. This edition will accordingly be found to differ considerably from the preceding. Alterations, more or less important, have been made on nearly every page. Among these may be mentioned the addition of Continued Fractions, the Extraction of the Roots of Numbers, Elimination by means of the Greatest Common Divisor, and a large collection of Miscellaneous Examples.

It is believed that this treatise contains as much of Algebra as can be profitably read in the time allotted to this study in most of our colleges, and that those subjects have been selected which are most important in a course of mathematical study. These materials I have endeavored to combine, so as to form a consistent treatise. I have aimed to cultivate in the mind of the student a habit of generalization, and to lead him to reduce every principle to its most general form. At the same time, I have been solicitous not to discourage the young beginner, who frequently finds it much more difficult to comprehend a general than a particular proposition. Accordingly, many of the Problems have been twice stated. I first give a simple numerical problem, and then repeat the same problem in a more general form. I have labored to develop, in a clear and intelligible manner, the most important properties of equations, and have bestowed great pains upon the selection of examples to illustrate these properties. Throughout the work I have endeavored to render the most important principles so prominent as to arrest attention, and I have reduced them, as far as practicable, to the form of concise and simple rules. It is believed that, in respect of difficulty, this treatise need not discourage any youth of fifteen years of age who possesses average abilities, while it is designed to form close habits of reasoning, and cultivate a truly philosophical spirit in more mature minds.

CONTENTS.

SECTION I.

DEFINITIONS AND NOTATION.

SECTION II.

SECTION III.

SECTION IV.

MULTIPLICATION.

SECTION V.

DIVISION.

SECTION VI.

FRACTIONS.

SECTION VII.

SIMPLE EQUATIONS.

SECTION VIII.

EQUATIONS WITH TWO OR MORE UNKNOWN QUANTITIES.

SECTION IX.

DISCUSSION OF EQUATIONS OF THE FIRST DEGREE.

SECTION X.

INVOLUTION AND POWERS.

SECTION XI.

EVOLUTION AND RADICAL QUANTITIES.

SECTION XII.

EQUATIONS OF THE SECOND DEGREE.

SECTION XIII.

RATIO AND PROPORTION.

SECTION XIV.

PROGRESSIONS.

SECTION XV.

GREATEST COMMON DIVISOR.—CONTINUED FRACTIONS.—PERMUTATIONS AND COMBINATIONS.

SECTION XVI.

INVOLUTION OF BINOMIALS.

SECTION XVII.

EVOLUTION OF POLYNOMIALS.

SECTION XVIII.

INFINITE SERIES.

SECTION XIX.

GENERAL THEORY OF EQUATIONS.

SECTION XX.

SOLUTION OF NUMERICAL EQUATIONS.

SECTION XXI.

LOGARITHMS.

ALGEBRA.

SECTION I.

PRELIMINARY DEFINITIONS AND NOTATION.

(*Article* 1.) WHATEVER is capable of increase or diminution, or will admit of mensuration, is called *magnitude* or *quantity.*

A sum of money, therefore, is a quantity, since we may increase it or diminish it. A line, a surface, a weight, and other things of this nature, are quantities; but an *idea* is *not* a quantity.

(2.) *Mathematics is the science of quantity,* or the science which investigates the means of measuring quantity. The operations of the mind, therefore, such as memory, imagination, judgment, &c., are not subjects of mathematical investigation, since they are not quantities.

(3.) Mathematics is divided into *pure* and *mixed.* Pure mathematics comprehends all inquiries into the relations of magnitude in the abstract, and without reference to material bodies. It embraces numerous subdivisions, such as Arithmetic, Algebra, Geometry, &c.

In the mixed mathematics these abstract principles are applied to various questions which occur in nature. Thus, in Surveying, the abstract principles of Geometry are applied to the measurement of land; in Navigation, the same principles are applied to the determination of a ship's place at sea; in Optics, they are employed to investigate the properties of light; and in Astronomy, to determine the distances of the heavenly bodies.

(4.) *Algebra is that branch of mathematics which enables us, by means of letters and other symbols, to abridge and generalize the reasoning employed in the solution of all questions relating to numbers.*

Arithmetic is the art or science of numbering. It treats of the nature and properties of numbers, but it is limited to certain methods of calculation which occur in common practice. Algebra is more comprehensive, and has been called by Newton, Universal Arithmetic.

(5.) The following are the main points of difference between Arithmetic and Algebra.

First, the operations of Algebra are more *general* than those of Arithmetic. In Arithmetic we represent quantities by particular numbers, as 2, 5, 7, &c., which numbers always retain the same value. The results obtained, therefore, are applicable only to the particular question proposed. Thus, if it is required to find the interest of a thousand dollars for three months at six per cent., the question may be solved by Arithmetic, and we obtain an answer, which is applicable only to this problem.

But in the solution of a general Algebraic problem we employ letters, to which any value may be attributed at pleasure. The results obtained, therefore, are equally applicable to all questions of a particular *class.* Thus, if we have given the sum and difference of two quantities, we may obtain by means of Algebra a general expression for the quantities themselves. This result will always be found true, whatever may be the magnitude of the quantities. Hence Algebra is adapted to the investigation of *general principles,* while Arithmetic is confined to operations upon *particular numbers.*

Secondly, Algebra enables us to solve a vast number of problems, which are *too difficult* for common Arithmetic. Some of the problems in Sections VII. and VIII. may be solved by Arithmetical methods; but others can not thus be resolved, particularly such problems as are given in Sections XII., XIV., &c.

Thirdly, in Arithmetic all the different quantities which enter into a problem are *blended together* in the result, so as to leave no trace of the operations to which they have been subjected. From a simple inspection of the result, we can not tell whether it was derived by multiplication or division, involution or evolution, or what connection it has with the given quantities of the problem. But in a general Algebraic solution, all the different quantities are preserved *distinct* from each other, and we see at a glance how all the data of the problem are combined in the result. Illustrations of this remark will be found in Section VII., &c.

Fourthly, the operations of Algebra are often far more *concise* than those of Arithmetic. Thus, although some of the problems in Sections VII. and VIII. may be solved Arithmetically, these solutions are generally much more tedious than the Algebraic. This advantage which is possessed by Algebra is partly due to the representation of the unknown quantities by letters, and their introduction into the operations as if they were already known, and partly to the fact that the operations of multiplication, division, &c., are at first merely *indicated*, and are not actually performed until an Algebraic expression has been reduced to its simplest form.

Finally, perhaps the most striking difference between Arithmetic and Algebra springs from the use of *negative quantities*, which give rise to many peculiar results.

The full purport of these remarks will be best apprehended after the student has made some progress in the study of Algebra.

(6.) A *definition* is the explanation of any term or word. It is essential to a perfect definition that it distinguish the thing defined from every thing else. Thus, if we say that man is a biped, it is an imperfect definition of man, because there are many other bipeds.

(7.) A *theorem* is the statement of some property, the truth of which is required to be proved. Thus the principle that the sum of the three angles of any triangle is equal to two right angles, is a theorem, the truth of which is demonstrated by Geometry.

(8.) A *problem* is a question requiring something to be done. Thus, to draw one line perpendicular to another is a problem. Theorems and problems are both known by the general term of *propositions*.

(9.) A *determinate* problem is one which admits of a certain or definite answer. An *indeterminate* problem commonly admits of an indefinite number of solutions; although when the answers are required in positive whole numbers, they are in some cases confined within certain limits, and in others the problem may be impossible.

(10.) The *solution* of a problem is the process by which we obtain the answer to it. A *numerical* solution is the obtaining an answer in numbers. A *geometrical* solution is the obtaining an answer by the principles of geometry. A *mechanical* solution is one which is gained by trials.

(11.) The principal symbols employed in Algebra are the following:

The sign + (an erect cross) is named *plus*, and is employed to denote the *addition* of two or more numbers. Thus, 5+3 signifies that we must add 3 to the number 5, in which case the result is 8. In the same manner, 11+6 is equal to 17; 14+10 is equal to 24, &c.

We also make use of the same sign to connect several numbers together. Thus, 7+5+9 signifies that to the number 7 we must add 5 and also 9, which make 21.

So, also, the sum of 8+5+13+11+1+3+10 is equal to 51.

(12.) In order to generalize numbers we represent them by letters, as a, b, c, d, &c. Thus the expression $a+b$ signifies the sum of two numbers, which we represent by a and b, and these may be any numbers whatever. In the same manner, $m+n+p+x$ signifies the sum of the numbers represented by these four letters. If we knew, therefore, the numbers represented by the letters, we could easily find by arithmetic the value of such expressions.

The first letters of the alphabet are commonly used to represent *known* quantities, and the last letters those which are *unknown*.

(13.) The sign − (a horizontal line) is called *minus*, and indicates that one quantity is to be *subtracted* from another. Thus, 8−5 signifies that the number 5 is to be taken from the number 8, which leaves a remainder of 3. In like manner, 12−7 is equal to 5, and 20−14 is equal to 6, &c.

Sometimes we may have several numbers to subtract from a single one. Thus, 16−5−4 signifies that 5 is to be subtracted from 16, and this remainder is to be further diminished by 4, leaving 7 for the result. In the same manner, 50−1−3−5−7−9 is equal to 25. So, also, $a-b$ signifies that the number designated by a is to be diminished by the number designated by b.

Quantities preceded by the sign + are called *positive* quantities; those preceded by the sign $-$, *negative* quantities. When no sign is prefixed to a quantity, $+$ is to be understood. Thus, $a+b-c$ is the same as $+a+b-c$.

(14.) The sign $\times$ (an inclined cross) is employed to denote the *multiplication* of two or more numbers. Thus, 3×5 signifies that 3 is to be multiplied by 5, making 15. In like manner, $a\times b$ signifies a multiplied by b; and $a\times b\times c$ signifies the continued product of the numbers designated by a, b, and c, and so on for any number of quantities.

Multiplication is also frequently indicated by placing a point between the successive letters. Thus, $a.b.c.d$ signifies the same thing as $a\times b\times c\times d$.

Generally, however, when numbers are represented by letters, their multiplication is indicated by writing them in succession without the interposition of any sign. Thus, $a\ b$ signifies the same thing as $a.b$ or $a\times b$; and $a\ b\ c\ d$ is equivalent to $a.b.c.d$, or $a\times b\times c\times d$.

It must be remarked that the notation $a.b$ or $a\ b$ is seldom employed except when the numbers are designated by letters. If, for example, we attempt to represent the product of the numbers 5 and 6 in this manner, 5 . 6 might be confounded with an integer followed by a decimal fraction; and 56 would be read *fifty-six*, according to the common system of notation.

The multiplication of numbers may, however, be expressed by placing a point between them, in cases where *no ambiguity* can arise from the use of this symbol. Thus, 1 . 2 . 3 . 4 is sometimes used to represent the continued product of the numbers 1, 2, 3, 4.

(15.) When two or more quantities are multiplied together, each of them is called a *factor*. Thus, in the expression 7×5, 7 is a factor, and so is 5. In the product abc there are three factors, a, b, c.

When a quantity is represented by a letter, it is called a *literal* factor, to distinguish it from a *numerical* factor, which is represented by an Arabic numeral. Thus, in the expression $5ab$, 5 is a numerical factor, while a and b are literal factors.

(16.) The character $\div$ (a horizontal line with a point above and below) shows that the quantity which precedes it is to be divided by that which follows.

Thus, $24 \div 6$ signifies that 24 is to be divided by 6, making 4. So, also, $a \div b$ is a divided by b.

Generally, however, the division of two numbers is indicated by writing the dividend above the divisor, and drawing a line between them.

Thus, $24 \div 6$ and $a \div b$ are usually written $\frac{24}{6}$ and $\frac{a}{b}$.

(17.) The sign $=$ (two horizontal lines) when placed between two quantities, denotes that they are equal to each other.

Thus, $7+6=13$ signifies that the sum of 7 and 6 is equal to 13. So, also, \$1=100 cents, is read one dollar equals one hundred cents; 3 shillings=36 pence, is read three shillings are equal to thirty-six pence. In like manner, $a=b$ signifies that a is equal to b; and $a+b=c-d$ signifies that the sum of the numbers designated by a and b is equal to the difference of the numbers designated by c and d.

(18.) The symbol $>$ is called the sign of *inequality*, and when placed between two numbers, denotes that one of them is *greater* than the other, the opening of the sign being turned toward the greater number.

Thus, $3<5$ signifies that 3 is less than 5, and $11>6$ denotes that 11 is greater than 6. So, also, $a>b$ shows that a is greater than b, and $c<d$ shows that c is less than d.

(19.) The *coefficient* of a quantity is the number or letter prefixed to it, showing how often the quantity is to be taken.

Thus, instead of writing $a+a+a+a+a$, which represents five a's added together, we write $5a$, where 5 is the coefficient of a. In like manner, $10ab$ signifies ten times the product of a and b. The coefficient may be either a whole number or a fraction. Thus, $\frac{3}{4}a$ signifies three fourths of a. When no coefficient is expressed, 1 is always to be understood. Thus, $1a$ and a signify the same thing.

The coefficient may be a *letter* as well as a figure. In the expression mx, m may be considered as the coefficient of x, because x is to be taken as many times as there are units in m. If m stands for 5, then mx is 5 times x.

In $4ax$, 4 may be considered as the coefficient of ax, or $4a$ may be considered as the coefficient of x.

(20.) The products formed by the successive multiplication of the same number by itself are called the *powers* of that number

Thus, $2\times2=4$, the second power of 2.
$2\times2\times2=8$, the third power.
$2\times2\times2\times2=16$, the fourth power, &c.
So, also, $3\times3=9$, the second power of 3.
$3\times3\times3=27$, the third power, &c.
Also, $a\times a=aa$, the second power of a.
$a\times a\times a=aaa$, the third power, &c.

In general, any power of a quantity is designated by the number of factors which form the product.

(21.) For the sake of brevity, powers are usually expressed by writing the root once, with a number above it at the right hand, showing how many times the root is taken as a factor. This number is called the *exponent* of the power.

Thus, instead of

aa, we write a^2, where 2 is the exponent of the power.
aaa, " a^3, where 3 is the exponent of the power.
$aaaa$, " a^4, where 4 is the exponent of the power.
$aaaaa$, " a^5, where 5 is the exponent of the power, &c.

When no exponent is expressed, 1 is always understood. Thus, a^1 and a signify the same thing.

Exponents may be attached to figures as well as letters. Thus, the product of 3 by 3 may be written 3^2, which equals 9

"	$3\times3\times3$	"	3^3,	"	27
"	$3\times3\times3\times3$	"	3^4,	"	81
"	$3\times3\times3\times3\times3$	"	3^5,	"	243

(22.) A *root* of a quantity is a factor, which, multiplied by itself a certain number of times, will produce the given quantity.

The symbol $\sqrt{\quad}$ is called the *radical sign*, and when prefixed to a quantity denotes that its root is to be extracted Thus,

$\sqrt[2]{9}$, or simply $\sqrt{9}$, denotes the square root of 9, which is 3.
$\sqrt[3]{64}$ denotes the cube root of 64, which is 4.
$\sqrt[4]{16}$ denotes the fourth root of 16, which is 2.

So, also,

$\sqrt[2]{a}$, or simply $\sqrt{a}$, is the square root of a.
$\sqrt[3]{a}$ denotes the third or cube root of a.
$\sqrt[4]{a}$ denotes the fourth root of a.
$\sqrt[n]{a}$ denotes the nth root of a, where n may represent any number whatever.

The number placed over the radical sign is called the *index* of the root. Thus, 2 is the index of the square root, 3 of the cube root, 4 of the fourth root, and n of the nth root. The index of the square root is usually omitted. Thus, instead of $\sqrt[2]{ab}$, we usually write $\sqrt{ab}$.

(23.) When four quantities are *proportional*, the proportion is expressed by points, as in arithmetic. Thus, $a:b::c:d$, signifies that a has to b the same ratio which c has to d.

(24.) A *vinculum* ——, or a *parenthesis* ( ), indicates that several quantities are to be subjected to the *same operation.*

Thus, $\overline{a+b+c}\times d$, or $(a+b+c)\times d$, denotes that the sum of a, b, and c is to be multiplied by d. But $a+b+c\times d$, denotes that c *only* is to be multiplied by d.

When the parenthesis is used, the sign of multiplication is generally omitted. Thus, $(a+b+c)\times d$ is the same as $(a+b+c)d$, or $d(a+b+c)$.

(25.) Three dots $\therefore$ are sometimes employed to denote *therefore* or *consequently.*

A few other symbols are employed in algebra, in addition to those here enumerated, which will be explained as they occur.

(26.) Every number written in algebraic language, that is, by aid of algebraic symbols, is called an *algebraic quantity*, or an *algebraic expression.*

Thus, $3a$ is the algebraic expression for three times the number a.

$4a^2$ is the algebraic expression for four times the square of the number a.

$7a^3b^4$ is the algebraic expression for seven times the third power of a multiplied by the fourth power of b.

(27.) An algebraic quantity, not composed of parts which are separated from each other by the sign of addition or subtraction, is called a *monomial*, or a quantity of one term, or simply a *term.*

Thus, $2a$, $5bc$, and $7xy^2$ are monomials.

(28.) An algebraic expression, which is composed of several terms, is called a *polynomial.*

Thus, $a+2b+5c-d$ is a polynomial.

A polynomial consisting of *two terms* only, is usually called a *binomial;* one consisting of *three terms* is called a *trinomial.*

Thus, $3a+5b$ is a binomial, and $a+3bc+xy$ is a trinomial.

(29.) The *numerical value* of an algebraic expression is the result obtained when we attribute particular values to the letters.

Suppose the expression is $2a^2b$.

If we make $a=2$ and $b=3$, the value of this expression will be $2\times2\times2\times3=24$.

If we make $a=4$ and $b=3$, the value of the same expression will be $2\times4\times4\times3=96$.

The numerical value of a polynomial is not affected by *changing the order* of the terms, provided we preserve their respective signs.

The expressions $a^2+2ab+b^2$, a^2+b^2+2ab, $b^2+2ab+a^2$, have all the same numerical value.

Thus, if $a=5$ and $b=2$, the value of a^2 will be 25, that of $2ab$ will be 20, and b^2 will be 4; and if these numbers are added together, their sum will be the same in whatever order they are placed. Thus,

25	25	20	20	4	4
20	4	25	4	25	20
4	20	4	25	20	25
49	49	49	49	49	49

(30.) Each of the literal factors which compose a term is called a *dimension* of this term; and the *degree* of a term is the *number* of these factors or dimensions. A numerical coefficient is not counted as a dimension.

Thus, $3a$ is a term of one dimension, or of the first degree.

$5ab$ is a term of two dimensions, or of the second degree.

$6a^2bc^3$ is a term of six dimensions, or of the sixth degree.

In general, the degree, or the number of dimensions of a term, is equal to the sum of the exponents of the letters contained in the term.

Thus, the degree of the term $5ab^2cd^3$ is $1+2+1+3$ or 7; that is, this term is of the seventh degree.

(31.) A polynomial is said to be *homogeneous* when all its terms are of the *same degree.*

Thus, $2a-3b+c$, is of the first degree and homogeneous.

$3a^2-4ab+b^2$, is of the second degree and homogeneous.

$2a^3+3a^2c-4c^2d$, is of the third degree and homogeneous.

$5a^3-2ab+c$, is *not* homogeneous

(32.) *Like* or *similar* terms are terms composed of the *same letters* affected with the *same exponents.*

Thus, $3ab$ and $7ab$ are similar terms.

$5a^2c$ and $3a^2c$ are also similar terms.

But $3ab^2$ and $4a^2b$ are *not* similar; for, although they contain the same letters, the same letters are not affected with the same exponents.

(33.) The *reciprocal* of a quantity is the quotient arising from dividing a unit by that quantity.

Thus, the reciprocal of 2 is $\frac{1}{2}$; the reciprocal of a is $\frac{1}{a}$.

(34.) A few examples are here subjoined, to exercise the learner on the preceding definitions and remarks.

Examples in which words are to be converted into algebraic symbols.

Ex. 1. What is the algebraic expression for the following statement? The second power of a, increased by twice the product of a and b, diminished by c, and increased by d, is equal to seventeen times f.

Ans. $a^2+2ab-c+d=17f.$

Ex. 2. The quotient of three divided by the sum of x and four, is equal to twice b diminished by eight.

Ex. 3. One third of the difference between six times x and four, is equal to the quotient of five divided by the sum of a and b.

Ex. 4. Three quarters of x increased by five, is equal to three sevenths of b diminished by seventeen.

Ex. 5. One ninth of the sum of six times x and five, added to one third of the sum of twice x and four, is equal to the product of a, b, and c.

Ex. 6. The quotient arising from dividing the sum of a and b by the product of c and d, is equal to four times the sum of e, f, g, and h.

(35.) *Examples in which the algebraic signs are to be translated into common language.*

Ex. 1. $\frac{x+a}{b}+\frac{x}{c}=\frac{d}{a+b}.$

Ans. The quotient arising from dividing the sum of x and a by b, increased by the quotient of x divided by c, is equal to the quotient of d divided by the sum of a and b.

Ex. 2. $7a^2+(b-c)\times(d+e)=g+h.$

How should the preceding example be read, when the first parenthesis is omitted?

Ex. 3. $\frac{a+g}{3+b-c}+\frac{6-4m}{3}=\frac{h}{7+a}$.

Ex. 4. $4\sqrt{ab}-25=\frac{2b+c}{3a-d}$.

Ex. 5. $2a\sqrt{b^2}-ac=5(h+d+x)$.

Ex. 6. $\frac{\sqrt{5b}+3\sqrt{c}}{2a+1}=5x+\frac{1}{4}$.

(36.) Find the value of the following expressions, when $a=6$, $b=5$, and $c=4$.

Ex. 1. $a^2+3ab-c^2$.

Ans. $36+90-16=110$.

Ex. 2. $a^2\times(a+b)-2abc$.

Ans. 156.

Ex. 3. $\frac{a^3}{a+3c}+c^2$.

Ans. 28.

Ex. 4. $c+\frac{2bc}{\sqrt{2ac+c^2}}$

Ex. 5. $\sqrt{b^2-ac}+\sqrt{2ac+c^2}$.

Ex. 6. $3\sqrt{c}+2a\sqrt{2a+b+2c}$.

Ex. 7. $(3\sqrt{c}+2a)\sqrt{2a+b+2c}$.

Ex. 8. In the expression $\frac{ax^2+b^2}{bx-a^2-c}$, let $a=3$, $b=5$, $c=2$, and $x=6$; what is its numerical value?

Ex. 9. What is the value of $\sqrt{5\sqrt{x}+5\sqrt{y}}+\sqrt{x}+\sqrt{y}$, when $x=9$ and $y=4$?

Ex. 10. What is the value of $x^4-4x^3+7x^2-6x$, when $x=3$?

Ex. 11. What is the value of $5(x^2+y^2)+4xy$, when $x=4$ and $y=6$?

Ex. 12. What is the value of $\sqrt{10+x}-\sqrt[3]{10+x}$, when $x=6$?

Ex. 13. What is the value of $2x^2+\sqrt{2x^2+1}$, when $x=2$?

Ex. 14. What is the value of $2x-7\sqrt{x}$, when $x=81$?

SECTION II.

ADDITION.

(37) Addition is the connecting of quantities together by means of their proper signs, and incorporating such as can be united into one sum.

It is convenient to distinguish three Cases.

CASE I.

When the quantities are similar and have the same signs.

RULE.

Add the coefficients of the several quantities together, and to their sum annex the common letter or letters, prefixing the common sign.

Thus, the sum of $3a$ and $5a$ is obviously $8a$. So, also, $-3a$ and $-5a$ make $-8a$; for the minus sign before each of the terms shows that they are to be subtracted, not from each other, but from some quantity which is not here expressed; and if $3a$ and $5a$ are to be successively subtracted from the same quantity, it is the same as subtracting at once $8a$.

EXAMPLES.

$3a$	$-3ab$	$2b+3x$	$a-2x^2$	$2a+y^2$
$5a$	$-6ab$	$5b+7x$	$4a-3x^2$	$5a+2y^2$
$7a$	$-ab$	$b+2x$	$3a-5x^2$	$9a+3y^2$
a	$-7ab$	$4b+3x$	$7a-x^2$	$4a+6y^2$
$16a$	$-17ab$			

The learner must continually bear in mind the remark of *Art.* 13, that when no sign is prefixed to a quantity, plus is always to be understood.

CASE II.

(38.) When the quantities are similar, but have different signs.

RULE.

Add all the positive coefficients together, and also all those that are negative; subtract the least of these results from the greater; to the difference annex the common letter or letters, and prefix the sign of the greater sum.

Thus, instead of $7a-4a$, we may write $3a$, since these two expressions obviously have the same value.

Also, if we have $5a-2a+3a-a$, this signifies that from $5a$ we are to subtract $2a$, add $3a$ to the remainder, and then subtract a from this last sum, the result of which operation is $5a$. But it is generally most convenient to take the sum of the positive quantities, which in this case is $8a$; then take the sum of the negative quantities, which in this case is $3a$; and we have $8a-3a$ or $5a$, the same result as before.

EXAMPLES.

$-3a$	$6x+5ay$	$2ay-7$	$-2a^2x$	$-6a^2+2b$
$+7a$	$-3x+2ay$	$-ay+8$	a^2x	$2a^2-3b$
$+8a$	$x-6ay$	$2ay-9$	$-3a^2x$	$-5a^2-8b$
$-a$	$2x+ay$	$3ay-11$	$7a^2x$	$4a^2-2b$
$+11a$	$6x+2ay$			

CASE III.

(39.) When some of the quantities are dissimilar.

RULE.

Collect all the like quantities together, by taking their sums or differences as in the two former cases, and set down those that are unlike, one after the other, with their proper signs.

Unlike quantities can not be united in one term. Thus, $2a$ and $3b$ neither make $5a$ nor $5b$. Their sum can only be written $2a+3b$.

EXAMPLES.

$2xy-2x^2$	$3x^2y+2ax$	$2ax-220$	$2x\ -18y$
$3x^2+xy$	$-2xy^2-ax^2$	$3x^2-2ax$	$3xy+10x$
x^2-xy	$-3y^2x+3ax^2$	$5x^2-3x$	$2x^2y+25y$
$4x^2-3xy$	$-8x^2y-ax$	$3x+100$	$12x^2y-xy$
$6x^2-xy$		$8x^2-120$	

(40.) When several quantities are to be added together, it is most convenient to *write all the similar terms under each other*, as in the following example.

Ex. 1. Add together

$$\begin{array}{l} 11bc+4ad-8ac+5cd \\ 8ac+7bc-2ad+4mn \\ 2cd-3ab+5ac+an \\ 9an-2bc-2ad+5cd \end{array}$$

These terms may be written thus:

$$\begin{array}{lllllll} 11bc & +4ad & -8ac & +5cd & +an & +4mn & -3ab \\ 7bc & -2ad & +8ac & +2cd & +9an & & \\ -2bc & -2ad & +5ac & +5cd & & & \\ \hline \text{Sum}\ 16bc & & +5ac & +12cd & +10an & +4mn & -3ab. \end{array}$$

Ex. 2. Add together the quantities

$$\begin{array}{l} 7m+3n-14p \\ 3a+9n-11m \\ 5p-4m+8n \\ 11n-2b-m \end{array}$$

Ans. $3a-2b-9m+31n-9p$.

Ex. 3. Add together

$$\begin{array}{l} 4a^2b+3c^3d-9m^2n-6ab^2 \\ 4m^2n-ab^2+5c^3d+7a^2b \\ 6m^2n-5c^3d+4mn^2-8ab^2 \\ 7mn^2+6c^3d-5m^2n-6a^2b \\ 9c^3d-10ab^2-8m^2n+12a^2b \end{array}$$

Ans. $17a^2b+18c^3d-12m^2n-25ab^2+11mn^2$.

Ex. 4. Add together

$$\begin{array}{l} 3b-a-6c-115d-9f \\ 6c-5f-d+6e-3a \\ 3a-2b-3c+27e+11f \\ 3e-7f+5b-8c+9d \\ 17c-6b-7a-2d-5e \end{array}$$

Ans. $-8a+6c-109d+31e-10f$.

Ex. 5. Add together

$$\begin{array}{l} 2ab^2+3ac^2+\ 9b^2x-\ 8hy^2+10ky \\ 2ab^2-3x^2\ -\ \ b^2x-\ 4ky^2-15hy \\ 5ky\ -\ hy^2-22ac^2-10x^2\ -\ 4ab^2 \\ 19ac^2-8b^2x+\ 9x^2\ +\ 6hy+\ 2ky^2 \end{array}$$

Ans. $-9hy^2+15ky-2ky^2-9hy-4x^2$.

(41.) It must be observed that the term addition is used in a more extended sense in algebra than in arithmetic. In arithmetic, where all quantities are regarded as positive, addition implies *augmentation.* The sum of two quantities will therefore be numerically greater than either quantity. Thus the sum of 7 and 5 is 12, which is numerically greater than either 5 or 7.

But in algebra we consider negative as well as positive quantities; and by the sum of two quantities, we mean their aggregate, regard being paid to their signs. Thus the sum of +7 and −5 is +2, which is numerically less than either 7 or 5. So, also, the sum of $+a$ and $-b$ is $a-b$. In this case, the algebraic sum is numerically the *difference* of the two quantities.

This is one instance, among many, in which the same terms are used in a much more general sense in the higher mathematics than they are in arithmetic.

Ex. 6. What is the sum of $3a^2b^2c^2+5a^3b^2c-7ax+2ab^2c^3-6ab^2c^3-9ax-4ab^2c^3-11a^2b^2c^2-17a^3b^2c+6ax$.

Ans.

Ex. 7. What is the sum of $9a^2-17a^3x+5a^2b^2c^2+4a^3x+8a^2-8a^2b^2c^2+3a^2+19a^2b^2c^2-2a^3x-11a^2+4a^2b^2c^2$.

Ans.

Ex. 8. What is the sum of $5am^3+13n-6xy^2+4ax+6n-2am^3+17ax+9xy^2+3am^3-11ax+7n-8xy^2$.

Ans.

Ex. 9. What is the sum of $15a^3y-4a^2x+10am-13a^2x-9am+7a^3y+12am-6a^3y+11a^2x-5a^3y+14am-8a^2x$.

Ans.

Ex. 10. What is the sum of $21am^3-3a^2b+11xy^2+15ag-5a^2b+13ag-6am^3+17a^2b-8xy^2-9ag+19a^2b-7xy^2$.

Ans.

SECTION III.

SUBTRACTION.

(42.) Subtraction is the taking of one quantity from another; or it is finding the difference between two quantities or sets of quantities.

Let it be required to subtract $8-3$ from 15.

Now $8-3$ is equal to 5.

And 5 subtracted from 15 leaves 10.

The result, then, must be 10. But, to perform the operation on the numbers as they were given, we first subtract 8 from 15, and obtain 7. This result is *too small* by 3, because the number 8 is *larger* by 3 than the number which was required to be subtracted. Therefore, in order to correct this result, the 3 must be added, and we have

$$15-8+3=10, \text{ as before.}$$

Again, let it be required to subtract $c-d$ from $a-b$. It is plain, that if the part c were alone to be subtracted, the remainder would be

$$a-b-c.$$

But as the quantity actually proposed to be subtracted is *less* than c by d, *too much* has been taken away by d, and, therefore, the true remainder will be greater than $a-b-c$ by d, and will hence be expressed by

$$a-b-c+d,$$

where the signs of the last two terms are both *contrary* to what they were given in the subtrahend.

(43.) Hence we deduce the following general

RULE.

Conceive the signs of all the terms of the subtrahend to be changed from + to −, or from − to +, and then collect the terms together, as in the several cases of addition.

It is better in practice to leave the signs of the subtrahend *unchanged*, and simply *conceive* them to be changed; that is, treat the quantities as if the signs were changed; for, otherwise, when we come to revise the work to detect any error in the operation, we might often be in doubt as to what were the signs of the quantities as originally proposed.

EXAMPLES.

From	$5a^2-2b$	$5xy+8x-2$	$10\ -8x-3xy$	$4ax-2x^2y$
Subtract	$2a^2+5b$	$3xy-8x-7$	$-x+3\ -\ xy$	$3ax-5xy^2$
Remainder	$3a^2-7b$		$7\ -7x-2xy$	

From $5a+4b-2c+7d$ From $11xy+2y^2-16x^2$
Take $3a+2b+\ c+5d$ Take $-\ 4xy+6y^2-18x^2$
Remainder $2a+2b-3c+2d$

From $6aby-4xy+4xz$ From $x^2+2xy+y^2$
Take $-3aby+5xz+3xy$ Take $x^2-2xy+y^2$
Remainder $9aby-\ xz-7xy$

From $3a^2+\ ax+2x^2-14a^2x+19ax^2-\ 4x^2\ \ +5a^2x^2$
Take $2a^2-4ax+\ x^2-15a^2x+11ax^2-15a^2x^2-4x^2$

Subtraction may be *proved* as in Arithmetic, by adding the remainder to the subtrahend. The sum should be equal to the minuend.

(44.) The term subtraction, it will be perceived, is used in a more general sense in algebra than in arithmetic. In arithmetic, where all quantities are regarded as positive, a number is always *diminished* by subtraction. But in algebra, the difference between two quantities may be numerically greater than either. Thus, the difference between $+a$ and $-b$ is $a+b$.

The distinction between positive and negative quantities may be illustrated by the scale of a thermometer. The degrees above zero are considered positive, and those below zero negative. From five degrees above zero to five degrees below zero, the numbers stand thus:

$$+5,\ +4,\ +3,\ +2,\ +1,\ 0,\ -1,\ -2,\ -3,\ -4,\ -5.$$

The difference between five degrees above zero and five degrees below zero is ten degrees, which is numerically the *sum* of the two quantities.

(45.) In practice, it is often sufficient merely to *indicate* the subtraction of a polynomial, without actually performing the operation. This is done by inclosing the polynomial in a parenthesis, and prefixing the sign $-$.

Thus, $5a-3b+4c-(3a-2b+8c)$

signifies that the entire quantity $3a-2b+8c$ is to be subtracted from $5a-3b+4c$. The subtraction is here merely *indicated.* If we actually perform the operation, the expression becomes

$$5a-3b+4c-3a+2b-8c$$

or $2a-b-4c$.

(46.) According to the preceding principle, polynomials may be written in a variety of forms.

Thus, $a-b-c+d$

is equivalent to $a-(b+c-d)$,

or to $a-b-(c-d)$,

or to $a+d-(b+c)$.

Transformations of this sort, which consist in decomposing a polynomial into two parts separated from each other by the sign $-$, are of frequent use in algebra. It is recommended to the student to write out polynomials like the above, containing both positive and negative terms, in all the possible modes, including several terms in a parenthesis.

In the following examples, let the results all be reduced to their simplest form.

Ex. 1. $a+b-(2a-3b)-(5a+7b)-(-13a+2b)=$.

Ex. 2. $37a-5f-(3a-2b-5c)-(6a-4b+3h)=$.

Ex. 3. $8a^2xy-5bx^2y+17cxy^2-9y^5-(a^2xy+3bx^2y-13cxy^2+20y^5)=$.

Ex. 4. $28ax^2-16a^2x^2+25a^2x-13a^4-(18ax^2+20a^2x^2-24a^2x-7a^4)=$.

(47.) It has already been remarked, in *Art.* 5, that algebra differs from arithmetic in the use of negative quantities, and it is important that the beginner should obtain clear ideas of their nature.

In many cases, the terms positive and negative are merely *relative.* They indicate some sort of *opposition* between two classes of quantities, such that if one class should be *added*, the other ought to be *subtracted.* Thus, if a ship sails alternately northward and southward, and the motion in one direction is called *positive*, the motion in the opposite direction should be considered *negative*.

Suppose a ship, setting out from the equator, sails northward 50 miles, then southward 27 miles, then northward 15 miles, then southward again 22 miles, and we wish to determine the last position of the ship. If we call the northerly motion +, the whole may be expressed algebraically thus:

$$+50-27+15-22,$$

which reduces to +16. The positive sign of the result indicates that the ship was 16 miles *north* of the equator.

Suppose the same ship sails again 8 miles north, then 35 miles south, the whole may be expressed thus:

$$+50-27+15-22+8-35,$$

which reduces to −11. The negative sign of the result indicates that the ship was now 11 miles *south* of the equator.

In this example we have considered the northerly motion +, and the southerly motion −; but we might with equal propriety have considered the southerly motion +, and the northerly motion −. It is, however, indispensable that we adhere to the *same system* throughout, and retain the proper sign of the result, as this sign shows whether the ship was at any time north or south of the equator.

In the same manner, if we consider easterly motion +, westerly motion must be regarded as −, and vice versa. And generally, when quantities which are estimated in different directions enter into the same algebraic expression, those which are measured in one direction being treated as +, those which are measured in the opposite direction must be regarded as −.

So, also, in estimating a man's property, *gains* and *losses* being of an opposite character, must be affected with different signs. Suppose a man, with a property of 1000 dollars, loses 300 dollars, afterward gains 100, and then loses again 400 dollars, the whole may be expressed algebraically thus:

$$+1000-300+100-400,$$

which reduces to +400. The + sign of the result indicates that he has now 400 dollars remaining in his possession. Suppose he further gains 50 dollars and then loses 700 dollars The whole may now be expressed thus:

$$+1000-300+100-400+50-700,$$

which reduces to -250. The $-$ sign of the result indicates that his losses exceed the sum of all his gains and the property originally in his possession; in other words, he owes 250 dollars more than he can pay, or, in common language, he is 250 dollars *worse than nothing.*

This phraseology must not be regarded as wholly figurative; for, in algebra, a negative quantity standing alone is regarded as less than nothing; and of two negative quantities, that which is *numerically the greatest* is considered as the *least;* for if from the same number we subtract successively numbers larger and larger, the remainders must continually *diminish.* Take any number, 5 for example, and from it subtract successively 1, 2, 3, 4, 5, 6, 7, 8, 9, &c., we obtain

$5-1$, $5-2$, $5-3$, $5-4$, $5-5$, $5-6$, $5-7$, $5-8$, $5-9$, &c., or reducing

$$4, 3, 2, 1, 0, -1, -2, -3, -4.$$

Whence we see that -1 should be regarded as smaller than nothing; -2 less than -1; -3 less than -2, &c.

EXAMPLES.

1. From $8x^2y^3-5xy+9a^2m$ take $6x^2y^3-10xy+7a^2m$.
Ans.

2. From $15a^3bx-19p^2q+11axy-25$ take $17a^3bx-13p^2q+9axy+7$. *Ans.*

3. From $10ax+19a^4x^3y-6ab^2+16gm$ take $4ax-13a^4x^3y+7ab^2+8gm$. *Ans.*

4. From $11a^2y-14amn+9gm+13a^5$ take $15a^2y+7amn-12gm-8a^5$. *Ans.*

5. From $13axy^2-6my+16a^2bc+8a^3$ take $9axy^2+2my+19a^2bc-4a^3$. *Ans.*

6. From $17a^2c-11bm+3xy^2+14amn^3$ take $15a^2c-21bm-6xy^2+8amn^3$. *Ans.*

7. From $35am^3-19bx+27y^4-11av^2$ take $15am^3-7bx+31y^4-23av^2$. *Ans.*

8. From $12ab^3-3cx-4xy^3-7abc^2$ take $8ab^3-6cx-5xy^3-12abc^2$. *Ans.*

9. From $40a^3+7b^2c-6b^3+3ax^2y^2$ take $12a^3-4b^2c+5b^3+4ax^2y^2$. *Ans.*

10. From $7ax^2-50amn-3by^3+6m$ take $9ax^2-2amn+3by^3-4m$. *Ans.*

SECTION IV.

MULTIPLICATION.

(48.) Multiplication is repeating the multiplicand as many times as there are units in the multiplier.

When several quantities are to be multiplied together, the result will be the same in whatever *order* the multiplication is performed.

This may be demonstrated in the following manner:

Let unity be repeated five times upon a horizontal line, and let there be formed four such parallel lines.

. . . . .
. . . .
. . . . .
. . . . .

Then it is plain that the number of units in the table is equal to the five units of the horizontal line, repeated as many times as there are units in a vertical column; that is, to the product of 5 by 4. But this sum is also equal to the four units of a vertical line repeated as many times as there are units in a horizontal line; that is, to the product of 4 by 5. Therefore, the product of 5 by 4 is equal to the product of 4 by 5. For the same reason, $2\times3\times4$ is equal to $2\times4\times3$, or $4\times3\times2$, or $3\times4\times2$, the product in each case being 24. So, also, if a, b, and c represent any three numbers, we shall have abc equal to bca or cab.

It is convenient to consider the subject of multiplication under three Cases.

CASE I.

(49.) *When both the factors are monomials.*

From *Article* 14, it appears that, in order to represent the multiplication of two monomials, such as $3abc$ and $5def$, we may write these quantities in succession without interposing any sign, and we shall have

$$3abc5def.$$

But, according to the principle stated in the preceding article, this result may be written

$$3\times5abcdef, \text{ or } 15abcdef.$$

Hence we deduce the following

RULE.

Multiply the coefficients of the two terms together, and to the product annex all the different letters in succession.

EXAMPLES.

Multiply	$12a$	$5a$	$7ab$	$7axy$	$6xyz$
By	$3b$	$6x$	$5ac$	$6ay$	ayz
Product	$36ab$				

From *Article* 48, it appears to be immaterial in what *order* the letters of a term are arranged; it is, however, generally most convenient to arrange them *alphabetically*.

(50.) We have seen in *Art.* 21, that when the same letter appears several times as a factor in a product, this is briefly expressed by means of an exponent. Thus, aaa is written a^3, the number 3 showing that a enters three times as a factor. Hence, if the same letters are found in two monomials which are to be multiplied together, the expression for the product may be abbreviated by adding the exponents of the same letters. Thus, if we are to multiply a^3 by a^2, we find a^3 equivalent to aaa, and a^2 to aa. Therefore the product will be $aaaaa$, which may be written a^5, a result which we might have obtained at once by adding together 3 and 2, the exponents of the common letter a.

Hence, since every factor of both multiplier and multiplicand must appear in the product, we have the following

RULE FOR THE EXPONENTS.

Powers of the same quantity may be multiplied by adding heir exponents.

EXAMPLES.

Multiply	$8a^2bc^2$	$2a^3b^2c$	$5a^4b^3c^2$	$2a^2b^3c^4$
By	$7abcd^2$	$8abc^3$	$7a^2b^4c^3d$	$5a^6bc^3$
Product	$56a^3b^2c^3d^2$			

CASE II.

(51.) *When the multiplicand is a polynomial.*

If $a+b$ is to be multiplied by c, this implies that the sum of the units in a and b is to be repeated c times; that is, the units in b repeated c times must be added to the units in a repeated also c times. Hence we deduce the following

RULE.

Multiply each term of the multiplicand separately by the multiplier, and add together the products.

EXAMPLES.

Multiply	$3a+2b$	a^2+2x+1	$3y^2+5xy+2$	$3x^3+xy+2y$
By	$4a$	$4x$	xy	$5x^2y$
Product	$12a^2+8ab$			

CASE III.

(52.) *When both the factors are polynomials.*

If $a+b$ is to be multiplied by $c+d$, this implies that the quantity $a+b$ is to be repeated as many times as there are units in the sum of c and d; that is, we are to multiply $a+b$ by c and d successively, and add the partial products. Hence we deduce the following

RULE.

Multiply each term of the multiplicand by each term of the multiplier separately, and add together the products.

EXAMPLES.

Multiply	$a+b$	$3x+2y$	$ax+b$	$3a+x$
By	$a+b$	$2x+3y$	$cx+d$	$2a+4x$
Product	$a^2+2ab+b^2$			

When several terms in the product are *similar*, it is most convenient to set them under each other, and then unite them by the rules for addition.

(53.) The examples thus far given in multiplication have been confined to *positive* quantities, and the products have all been positive. We must now establish a general rule for the signs of the product.

First, if $+a$ is to be multiplied by $+b$, this signifies that $+a$ is to be repeated as many times as there are units in b, and the result is $+ab$. That is, a plus quantity multiplied by a plus quantity gives a plus result.

Secondly, if $-a$ is to be multiplied by $+b$, this signifies that $-a$ is to be repeated as many times as there are units in b. Now $-a$ taken twice is obviously $-2a$, taken three times is $-3a$, &c.; hence, if $-a$ is repeated b times, it will make $-ba$ or $-ab$. That is, a minus quantity multiplied by a plus quantity gives minus.

Thirdly, to determine the sign of the product when the multiplier is a minus quantity, let it be proposed to multiply $8-5$ by $6-2$. By this we understand that the quantity $8-5$ is to be repeated as many times as there are units in $6-2$. If we multiply $8-5$ by 6, we obtain $48-30$; that is, we have repeated $8-5$ six times. But it was only required to repeat the multiplicand four times, or $(6-2)$. We must therefore *diminish* this product by twice $(8-5)$, which is $16-10$; and this subtraction is performed by changing the signs of the subtrahend; hence we have

$$48-30-16+10,$$

which is equal to 12. This result is obviously correct; for $8-5$ is equal to 3, and $6-2$ is equal to 4; that is, it was required to multiply 3 by 4, the result of which is 12, as found above.

In order to generalize this reasoning, let it be proposed to multiply $a-b$ by $c-d$.

If we multiply $a-b$ by c, we obtain $ac-bc$. But $a-b$ was only to be taken $c-d$ times; therefore, in this first operation, we have repeated it too many times by the quantity d. Hence, to have the true product, we must subtract d times $a-b$ from $ac-bc$. But d times $a-b$ is equal to $ad-bd$, which, subtracted from $ac-bc$, gives

$$ac-bc-ad+bd.$$

Thus we see that $+a$ multiplied by $-d$ gives $-ad$; and $-b$ multiplied by $-d$ gives $+bd$. Hence a plus quantity multiplied by a minus quantity gives minus, and a minus quantity multiplied by a minus quantity gives plus.

(54.) The preceding results may be briefly expressed as follows:

+ multiplied by +, and − multiplied by −, give +.

+ multiplied by −, and − multiplied by +, give −.

Or, the product of two quantities having the *same sign*, has the sign *plus;* the product of two quantities having *different signs*, has the sign *minus*.

(55.) The whole doctrine of multiplication is therefore comprehended in the following

RULE.

Multiply each term of the multiplicand by each term of the multiplier, and add together all the partial products, observing that like signs require + in the product, and unlike signs −.

EXAMPLE I.

Multiply	$5a^4-2a^3b+4a^2b^2$
By	$a^3-4a^2b+2b^3$
Partial Products	$5a^7-2a^6b+4a^5b^2$
	$-20a^6b+8a^5b^2-16a^4b^3$
	$+10a^4b^3-4a^3b^4+8a^2b^5$
Result	$5a^7-22a^6b+12a^5b^2-6a^4b^3-4a^3b^4+8a^2b^5$.

Ex. 2. Multiply $4a^3-5a^2b-8ab^2+2b^3$ by $2a^2-3ab-4b^2$.
Ans. $8a^5-22a^4b-17a^3b^2+48a^2b^3+26ab^4-8b^5$.

Ex. 3. Multiply $3a^2-5bd+ef$ by $-5a^2+4bd-8ef$.
Ans. $-15a^4+37a^2bd-29a^2ef-20b^2d^2+44bdef-8e^2f^2$

Ex. 4. Multiply $x^4+2x^3+3x^2+2x+1$ by x^2-2x+1.
Ans. x^6-2x^3+1.

Ex. 5. Multiply $14a^3e-6a^2bc+c^2$ by $14a^3e+6a^2bc-c^2$.

Ex. 6. Multiply $3a^3+35a^2b-17ab^2-13b^3$ by $3a^2+26ab-57b^2$.

(56.) Since in the multiplication of two monomials every factor of both quantities appears in the product, it is obvious that the *degree* of the product will be equal to the *sum* of the degrees of the multiplier and multiplicand. Hence, also, if two polynomials are *homogeneous*, their product will be homogeneous.

Thus, in the first of the preceding examples, all the terms of the multiplicand being of the fourth degree, and those of the multiplier of the third degree, all the terms of the product are of the seventh degree. For a like reason, in the second example, all the terms of the product are of the fifth degree; in the third example, they are of the fourth degree; and in the sixth example, they are of the fifth degree.

This remark will enable us to detect any error in the multiplication, so far as concerns the exponents. For example, if we find in one of the terms of a product which should be homogeneous, the sum of the exponents equal to 6, while in all the other terms it is equal to 7, a mistake has evidently been committed in the formation of one of the terms.

(57.) When the product arising from the multiplication of two polynomials does not admit of any reduction of similar terms, the *whole number* of terms in the product is equal to the number of terms in the multiplicand, multiplied by the number of terms in the multiplier.

Thus, if we have five terms in the multiplicand and four terms in the multiplier, the whole number of terms in the product will be 5×4, or 20. In general, if there be m terms in the multiplicand and n terms in the multiplier, the whole number of terms in the product will be $m\times n$.

(58.) If the product contains *similar terms*, the number of terms in the product when reduced may be much less; but it is important to observe, that among the different terms of the product there are always two which *can not be combined* with any others. These are,

1. The term arising from the multiplication of the two terms affected with the *highest* exponent of the same letter.

2. The term arising from the multiplication of the two terms ffected with the *lowest* exponent of the same letter.

For it is evident, from the rule of exponents, that these two partial products must involve the letter in question, the one with a *higher*, and the other with a *lower* exponent than any of the other partial products, and therefore can not be similar to any of them. Hence the product of two polynomials can never contain less than two terms.

(59.) For many purposes, it is sufficient merely to *indicate* the multiplication of two polynomials, without actually performing the operation. This is effected by inclosing the quantities in parentheses, and writing them in succession with or without the interposition of any sign.

Thus, $(a+b+c)$ $(d+e+f)$ signifies that the sum of a, b, and c is to be multiplied by the sum of d, e, and f.

When the multiplication is actually performed, the expression is said to be *expanded.*

(60.) The following Theorems are of such extensive application that they should be carefully committed to memory.

THEOREM I.

The square of the sum of two quantities is equal to the square of the first, plus twice the product of the first by the second, plus the square of the second.

Thus, if we multiply $a+b$
By $a+b$

$$\begin{array}{l} a^2+ab \\ \quad\;\; ab+b^2 \end{array}$$

We obtain the product $a^2+2ab+b^2$.

Hence, if we wish to obtain the square of a binomial, we can write out the terms of the result at once according to this theorem without the necessity of performing an actual multiplication.

EXAMPLES.

1. $(2a+b)^2=$.
2. $(a+3b)^2=$.
3. $(3a+3b)^2=$.
4. $(4a+3b)^2=$.
5. $(5a^2+b)^2=$.
6. $(5a^2+7ab)^2=$.
7. $(5a^3+b)^2=$.
8. $(5a^3+8a^2b)^2=$.
9. $(1+\frac{1}{3})^2=$.
10. $(3+\frac{1}{5})^2=$.

This theorem deserves particular attention, for one of the

most common mistakes of beginners is to call the square of $a+b$ equal to a^2+b^2.

THEOREM II.

(61.) *The square of the difference of two quantities is equal to the square of the first, minus twice the product of the first and second, plus the square of the second.*

Thus, if we multiply $a-b$

By $a-b$

$$\begin{array}{r} a^2-ab \\ -ab+b^2 \\ \hline a^2-2ab+b^2. \end{array}$$

We obtain the product $a^2-2ab+b^2$.

EXAMPLES.

1. $(a-2b)^2=.$
2. $(2a-3b)^2=.$
3. $(5a-4b)^2=.$
4. $(6a^2-x)^2=.$
5. $(6a^2-3x)^2=.$
6. $(7a^2-b)^2=.$
7. $(7a^2-12ab)^2=.$
8. $(7a^2b^2-12ab)^2=.$
9. $(2-\frac{1}{3})^2=.$
10. $(4-\frac{1}{5})^2=.$

Here, also, beginners often commit the mistake of putting the square of $a-b$ equal to a^2-b^2.

THEOREM III.

(62.) *The product of the sum and difference of two quantities is equal to the difference of their squares.*

Thus, if we multiply $a+b$

By $a-b$

$$\begin{array}{r} a^2+ab \\ -ab-b^2 \\ \hline a^2-b^2. \end{array}$$

We obtain the product a^2-b^2.

EXAMPLES.

1. $(2a+b)\ (2a-b)=.$
2. $(3a+4b)\ (3a-4b)=.$
3. $(7a+x)\ (7a-x)=.$
4. $(7ab+x)\ (7ab-x)=.$
5. $(8a+b)\ (8a-b)=.$
6. $(8a+7bc)\ (8a-7bc)=.$
7. $(5a^2+6b^3)\ (5a^2-6b^3)=$
8. $(5x^2y+3xy^2)\ (5x^2y-3xy^2)=.$

9. $(3+\frac{1}{4})\ (3-\frac{1}{4})=$.
10. $(4+\frac{1}{3})\ (4-\frac{1}{3})=$.

The student should be drilled upon examples like those appended to the preceding theorems until he can produce the results mentally with as great facility as he could read them if exhibited upon paper.

The utility of these theorems will be the more apparent, the more complicated the expressions to which they are applied. Frequent examples of their application will be seen hereafter.

(63.) The same theorems will enable us to resolve many complicated expressions into their factors.

1. Resolve $a^2+4ab+4b^2$ into its factors.

Ans. $(a+2b)\ (a+2b)$.

2. Resolve $a^2-6ab+9b^2$ into its factors.
3. Resolve $9a^2-24ab+16b^2$ into its factors.
4. Resolve a^4-b^4 into three factors.
5. Resolve a^6-b^6 into its factors.
6. Resolve a^8-b^8 into four factors.
7. Resolve $25a^4-60a^2b^3+36b^6$ into its factors.
8. Resolve n^2+2n+1 into its factors.
9. Resolve $4m^2n^2-4mn+1$ into its factors.
10. Resolve $49a^4b^4-168a^3b^3+144a^2b^2$ into its factors.
11. Resolve n^3+2n^2+n into three factors.
12. Resolve $1-\frac{1}{25}$ into two factors.
13. Resolve $4-\frac{1}{49}$ into two factors.

MULTIPLICATION BY DETACHED COEFFICIENTS.

(64.) The coefficients of a product depend simply upon the coefficients of the two factors, and not upon the literal parts of the terms. Hence we may obtain the coefficients of the product by multiplying the coefficients of the multiplicand severally by the coefficients of the multiplier. To these coefficients the proper letters may afterward be annexed. This will be best understood from a few examples.

Thus, take the first example of *Art.* 52, to multiply $a+b$ by $a+b$.

The coefficients of the multiplicand are	$1+1$
" " multiplier	$1+1$
	$1+1$
	$\quad 1+1$
Coefficients of the product	$1+2+1$
or, supplying the letters, we obtain	$a^2+2ab+b^2$,

which is the same result as before obtained.

Ex. 2. Multiply $3a^2+4ax-5x^2$ by $2a^2-6ax+4x^2$.

Coefficients of multiplicand	$3+\ 4-\ 5$
" multiplier	$2-\ 6+\ 4$
	$6+\ 8-10$
	$\quad -18-24+30$
	$\qquad +12+16-20$
Coefficients of the product	$6-10-22+46-20$

It may seem difficult in this case to supply the letters; but a little consideration will render it perfectly plain. Thus, $3a^2\times 2a^2$ is equal to $6a^4$; hence a^4 is the proper letter to be attached to the first coefficient. For the same reason, x^4 is the proper letter to be attached to the last coefficient. Moreover, we see that both the proposed polynomials are homogeneous, and of the second degree. Hence the product must be homogeneous, and of the fourth degree. The powers of a must decrease successively by unity, beginning with the first term, while those of x increase by unity. Hence the required product is

$$6a^4-10a^3x-22a^2x^2+46ax^3-20x^4.$$

Ex. 3. Multiply $x^3+x^2y+xy^2+y^3$ by $x-y$.

Ex. 4. Multiply x^3-3x^2+3x-1 by x^2-2x+1.

Ex. 5. Multiply $2a^3-3ab^2+5b^3$ by $2a-5b$.

If we should proceed with this example precisely in the same manner as with the preceding, we should commit an error by attempting to unite terms which are dissimilar. The reason is, that the multiplicand does not contain the usual complete series of powers of a. The term containing the second power of a is wanting. This does not render the method inapplicable, but it is necessary to preserve dissimilar terms distinct from each other; and since, while we are are operating on the coefficients, we have not the advantage of the letters to indicate what are similar terms, we supply the place of the defi cient term by a cipher. The operation will then proceed with entire regularity.

$$\begin{array}{l} 2+\ 0-3+\ 5 \\ \underline{2-\ 5\qquad\quad} \\ 4+\ 0-6+10 \\ \underline{\quad-10-0+15-25} \\ 4-10-6+25-25 \end{array}$$

Hence the product is

$$4a^4-10a^3b-6a^2b^2+25ab^3-25b^4.$$

Ex. 6. Multiply $2a^3-3ab^2+5b^3$ by $2a^2-5b^2$.

Here there is a term in each polynomial to be supplied by a cipher.

The preceding examples are intended to lead the student to consider the properties of coefficients by themselves, and prepare him for some investigations which are to follow, particularly in Section XX. The beginner, however, in attempting to apply the method, must be cautious not to unite dissimilar terms.

MISCELLANEOUS EXAMPLES.

1. Multiply $15a+8b-2c+7d$ by $4a-3b-5c+11d$.
Ans.
2. Multiply $3x^3-3x^2y^2+2y^3$ by $4x^3-3x^2y^2-6y^3$.
Ans.
3. Multiply $3x^2+8xy-5$ by $4x^2-7xy+9$.
Ans.
4. Multiply $6a^2-5ax-3x^2$ by $4a^2-3ax+2x^2$.
Ans.
5. Multiply $5x^2+xy+4y^2$ by $x^2-2xy+3y^2$.
Ans.
6. Multiply $x^3-x^2y+xy^2$ by $x^2-xy^2-y^3$.
Ans.
7. Multiply $7a^2x^2-3a^3x^3-5x^4$ by $4a^2x^2-2a^3x^3-3x^4$.
Ans.
8. Multiply $7a^4b^4-5a^3b^5+3b^6$ by $8a^3b^3+3a^2b-2b^3$.
Ans.
9. Multiply $6a^3-4ab^2-8c^3$ by $7a^3-5ab^2-3c^3$.
Ans.

SECTION V.

DIVISION.

(65.) The object of division in Algebra is the same as in Arithmetic, viz., *The product of two factors being given, and one of the factors, to find the other factor.*

The dividend is the product of the divisor and quotient, the divisor is the given factor, and the quotient is the factor required to be found.

CASE I.

(66.) *When the divisor and dividend are both monomials.*

Suppose we have 63 to be divided by 7. We must find such a factor as, multiplied by 7, will give exactly 63. We perceive that 9 is such a number, and therefore 9 is the quotient obtained when we divide 63 by 7.

Also, if we have to divide ab by a, it is evident that the quotient will be b; for a multiplied by b gives the dividend ab. So, also, $12mn$ divided by $3m$ gives $4n$; for $3m$ multiplied by $4n$ makes $12mn$.

Suppose we have a^5 to be divided by a^2. We must find a number which, multiplied by a^2, will produce a^5. We perceive that a^3 is such a number; for, according to *Art.* 50, we multiply a^3 by a^2, by adding the exponents 2 and 3, making 5. That is, the exponent 3 of the quotient is found by subtracting 2, the exponent of the divisor, from 5, the exponent of the dividend. Hence the following

RULE OF EXPONENTS IN DIVISION.

In order to divide quantities expressed by different powers of the same letter, *subtract the exponent of the divisor from the exponent of the dividend.*

EXAMPLES.

Divide	a^8	a^7	b^6	c^9	h^7	x^9	y^m
By	a^3	a^2	b^3	c^4	h^4	x^5	y^n
Quotient	a^5						

Let it be required to divide $35a^5$ by $5a^2$. We must find a quantity which, multiplied by $5a^2$, will produce $35a^5$. Such a quantity is $7a^3$; for, according to *Arts.* 49 and 50, $7a^3 \times 5a^2$ is equal to $35a^5$. Therefore, $35a^5$ divided by $5a^2$ gives for a quotient $7a^3$; that is, we have divided 35, the coefficient of the dividend, by 5, the coefficient of the divisor, and have subtracted the exponent of the divisor from the exponent of the dividend.

(67.) Hence, for the division of monomials, we have the following

RULE.

1. *Divide the coefficient of the dividend by the coefficient of the divisor.*

2. *Subtract the exponent of each letter in the divisor from the exponent of the same letter in the dividend.*

EXAMPLES

1. Divide $20x^3$ by $4x$. *Ans.* $5x^2$.
2. Divide $25a^3xy^4$ by $5ay^2$.
3. Divide $72ab^5x^2$ by $12b^3x$.
4. Divide $77a^3b^5c^6$ by $11ab^3c^4$.
5. Divide $272a^3b^4c^5x^6$ by $17a^2b^3cx^4$.
6. Divide $250x^7y^8z^3$ by $5xyz^3$.
7. Divide $48a^3b^3c^2d$ by $12ab^2c$.
8. Divide $150a^5b^8cd^3$ by $30a^3b^5d^2$.

(68.) The rule given in *Art.* 66 conducts, in some cases, to *negative exponents.*

Thus, let it be required to divide a^3 by a^5. We are directed to subtract the exponent of the divisor from the exponent of the dividend. We thus obtain

$$a^{3-5}=a^{-2}.$$

But a^3 divided by a^5 may be written $\frac{a^3}{a^5}$; and since the value of a fraction is not altered by dividing both numerator and denominator by the same quantity, this expression is equivalent to $\frac{1}{a^2}$.

Hence a^{-2} is the same as $\frac{1}{a^2}$,

and these expressions may be used indifferently for each other.

So, also, if a^2 is to be divided by a^5, this may be written

$$\frac{a^2}{a^5}=\frac{1}{a^3}=a^{-3}$$

In the same manner we find

$$a^{-m}=\frac{1}{a^m}.$$

That is, *the reciprocal of a quantity is equal to the same quantity with the sign of its exponent changed.*

So, also, $$\frac{a}{b^2c}=\frac{ab^{-2}}{c}=ab^{-2}c^{-1}.$$

And $$\frac{ad^{-4}}{b}=\frac{a}{bd^4}$$

(69.) Hence any factor may be transferred from the numerator to the denominator of a fraction, or from the denominator to the numerator, by *changing the sign of its exponent.*

Thus, $$\frac{a}{b}=ab^{-1}.$$

$$\frac{a^{-2}}{b^{-2}}=a^{-2}b^2.$$

$$\frac{a^4b^{-m}}{c^2d^{-3}}=a^4b^{-m}c^{-2}d^3.$$

That is, the denominator of a fraction may be entirely removed, and an *integral* form be given to any fractional expression.

This use of negative exponents must be understood simply as a convenient notation, and not as a method of actually destroying the denominator of a fraction. Still this new notation has many advantages, and is often employed, as will be seen hereafter.

When the division can not be exactly performed, it may be expressed in the form of a fraction, and this fraction may be reduced to its lowest terms, according to a method to be explained in *Art.* 83.

(70.) It frequently happens that the exponents of certain letters in the dividend are the *same* as in the divisor.

Let it be required to divide a^2 by a^2. The quotient is obviously 1, for every number is contained in itself once. But if we apply the rule of exponents, *Art.* 66, we shall have

$$a^{2-2} \text{ or } a^0.$$

Hence $$a^0=1.$$

Again, let it be required to divide a^m by a^m. The quotient is obviously 1, as before; and applying the rule of exponents, we obtain

$$a^{m-m} \text{ or } a^0.$$

That is, *every quantity affected with the exponent zero, is equal to unity.*

This notation has the advantage of preserving a trace of a letter which has disappeared in the operation of division. Thus, let it be required to divide a^3b^2 by a^2b^2. The quotient will be ab^0. This expression is of the same value as a alone, and is commonly so written. If, however, it was important to indicate that the letter b originally entered into the expression, this might be done without at all affecting the value of the result by writing it

$$ab^0.$$

(71.) The proper *sign* to be prefixed to a quotient is readily deduced from the principles already established for multiplication. The product of the divisor and quotient must be equal to the dividend. Hence,

$$\text{because } \left.\begin{array}{r} +a\times+b=+ab \\ -a\times+b=-ab \\ +a\times-b=-ab \\ -a\times-b=+ab \end{array}\right\} \text{ therefore } \left\{\begin{array}{l} +ab\div+b=+a. \\ -ab\div+b=-a. \\ -ab\div-b=+a. \\ +ab\div-b=-a. \end{array}\right.$$

Hence we have the following

RULE FOR THE SIGNS.

When both the dividend and divisor have the same sign, the quotient will have the sign +; when they have different signs, the quotient will have the sign −.

EXAMPLES.

Ex. 1. Divide $-15ay^2$ by $3ay$.

Ex. 2. Divide $-18ax^2y$ by $-9ax$.

Ex. 3. Divide $150a^3bc$ by $-5ac$.

Ex. 4. Divide $40a^3b^4c$ by $-abc$.

CASE II.

(72.) *When the divisor is a monomial, and the dividend a polynomial.*

We have seen, *Art.* 51, that when a single term is multiplied into a polynomial, the former enters into *every term* of the latter.

Thus, $a(a+b)=a^2+ab$.

Hence $(a^2+ab)\div a=a+b$.

Whence we deduce the following

RULE.

Divide each term of the dividend by the divisor, as in the former case.

EXAMPLES.

Ex. 1. Divide $3x^3+6x^2+3ax-15x$ by $3x$.

Ans. $x^2+2x+a-5$.

Ex. 2. Divide $3abc+12abx-9a^2b$ by $3ab$.

Ex. 3. Divide $40a^3b^3+60a^2b^2-17ab$ by $-ab$.

Ex. 4. Divide $15a^2bc-10acx^2+5ac^2d^2$ by $-5a^2c$.

Ex. 5. Divide $6a^2x^4y^6-12a^3x^3y^6+15a^4x^5y^3$ by $3a^2x^2y^2$.

Ex. 6. Divide $x^{n+1}-x^{n+2}+x^{n+3}-x^{n+4}$ by x^n.

Ex. 7. Divide $12a^4y^6-16a^5y^5+20a^6y^4-28a^7y^3$ by $-4a^4y^3$.

CASE III.

(73.) *When the divisor and dividend are both polynomials.*

Let it be required to divide $2ab+a^2+b^2$ by $a+b$.

The object of this operation is to find a third polynomial which, multiplied by the second, will reproduce the first.

It is evident that the dividend is composed of all the partial products arising from the multiplication of each term of the divisor by each term of the quotient, these products being added together and reduced. Hence, if we can discover a term of the dividend which is derived *without reduction* from the multiplication of a term of the divisor by a term of the quotient, then dividing this term by the corresponding term of the divisor, we shall be sure to obtain a term of the quotient.

But from *Art.* 58, it appears that the term a^2, which contains the *highest exponent* of the letter a, is derived, *without reduction*, from the multiplication of the two terms of the divisor and quotient which are affected with the highest exponent of the same letter. Dividing then the term a^2 by the term a of the divisor, we obtain a, which we are certain must be one term of the quotient sought. Multiplying each term of the divisor by a, and subtracting this product from the proposed dividend, the remainder may be regarded as the product of the divisor by the remaining terms of the quotient. We shall then obtain another term of the quotient by dividing that term of the remainder affected with the highest exponent of a, by the term a of the divisor, and so on.

Thus we perceive that at each step we are obliged to search for that term of the dividend which is affected with the highest exponent of one of the letters, and divide it by that term of the divisor which is affected with the highest exponent of the same letter. We may avoid the necessity of *searching* for this term by arranging the terms of the divisor and dividend *in the order of the powers of one of the letters.*

The operation will then proceed as follows:

$$
\begin{array}{r|l}
\text{The arranged dividend } = a^2+2ab+b^2 & a+b = \text{ the divisor.} \\
\underline{a^2+\ ab\qquad} & a+b = \text{ the quotient.} \\
ab+b^2 & = \text{ first remainder.} \\
\underline{ab+b^2} & \\
0 &
\end{array}
$$

It is generally convenient in Algebra to place the divisor on the right of the dividend, and the quotient directly under the divisor.

(74.) From this investigation we deduce the following

RULE FOR THE DIVISION OF POLYNOMIALS.

1. *Arrange the dividend and divisor according to the powers of the same letter*

2. *Divide the first term of the dividend by the first term of the divisor, the result will be the first term of the quotient.*

3. *Multiply the divisor by this term, and subtract the product from the dividend.*

4. *Divide the first term of the remainder by the first term of the divisor, the result will be the second term of the quotient.*

5. *Multiply the divisor by this term, and subtract the product from the last remainder. Continue the same operation till all the terms of the dividend are exhausted.*

If the divisor is not exactly contained in the dividend, the quantity which remains after the division is finished must be placed over the divisor in the form of a fraction, and annexed to the quotient.

EXAMPLES.

1. Divide $2a^2b+b^3+2ab^2+a^3$ by a^2+b^2+ab.

Ans. $a+b$.

2. Divide $x^3-a^3+3a^2x-3ax^2$ by $x-a$.

Ans. $x^2-2ax+a^2$.

3. Divide $a^6+z^6+2a^3z^3$ by a^2-az+z^2.

Ans. $a^4+a^3z+az^3+z^4$.

4. Divide $a^6-16a^3x^3+64x^6$ by $a^2-4ax+4x^2$.

5. Divide $a^4+6a^2x^2-4a^3x+x^4-4ax^3$ by $a^2-2ax+x^2$.

Ans. $a^2-2ax+x^2$.

6. Divide $x^4+x^2y^2+y^4$ by x^2+xy+y^2.

7. Divide $12x^4-192$ by $3x-6$.

Ans. $4x^3+8x^2+16x+32$

8. Divide $6x^6-6y^6$ by $2x^2-2y^2$.

9. Divide $a^6+3a^2b^4-3a^4b^2-b^6$ by $a^3-3a^2b+3ab^2-b^3$.

Ans. $a^3+3a^2b+3ab^2+b^3$.

10. Divide a^3-b^3 by $a-b$.

11. Divide a^4-b^4 by $a-b$.

If the first term of the arranged dividend is not divisible by the first term of the arranged divisor, the complete division is *impossible.*

(75.) Hitherto we have supposed the terms of the quotient to be obtained by dividing that term of the dividend affected with the *highest* exponent of a certain letter. But, from the second remark of *Art.* 58, it appears that the term of the dividend affected with the *lowest* exponent of any letter is derived, without reduction, from the multiplication of a term of the divisor by a term of the quotient. Hence we may obtain a term of the quotient by dividing the term of the dividend affected with the lowest exponent of any letter, by the term of the divisor containing the lowest power of the same letter, and nothing prevents our operating upon the highest and lowest exponents of a certain letter alternately in the same example.

(76.) From the examples of *Art.* 74, we perceive that a^3-b^3 is divisible by $a-b$; and a^4-b^4 is divisible by $a-b$. We shall find the same to hold true, whatever may be the value of the exponents of the two letters. That is, *the difference of any two powers of the same degree is divisible by the difference of their roots.*

Thus, let us divide a^5-b^5 by $a-b$.

$$\begin{array}{r|l} a^5-b^5 & a-b \\ \hline a^5-a^4b & a^4 \\ \hline a^4b-b^5. & \end{array}$$

The first term of the quotient is a^4, and the first remainder is a^4b-b^5, which may be written

$$b\,(a^4-b^4).$$

Now if, after a division has been partially performed, the remainder is divisible by the divisor, it is obvious that the dividend is completely divisible by the divisor. But we have already found that a^4-b^4 is divisible by $a-b$; therefore a^5-b^5 is also divisible by $a-b$; and in the same manner it may be proved that a^6-b^6 is divisible by $a-b$, and so on.

To exhibit this reasoning in a more general form, let us represent any exponent whatever by the letter n, and let us divide a^n-b^n by $a-b$.

$$\begin{array}{lr|l} & a^n-b^n & a-b \\ \hline & a^n-ba^{n-1} & a^{n-1} \\ \hline \text{First remainder} = & ba^{n-1}-b^n. & \end{array}$$

Dividing a^n by a, we have, by the rule of exponents, a^{n-1} for the quotient. Multiplying $a-b$ by this quantity, and subtracting the product from the dividend, we have for the first remainder $ba^{n-1}-b^n$, which may be written

$$b\,(a^{n-1}-b^{n-1}).$$

Now if this remainder is divisible by $a-b$, it is obvious that the dividend is divisible by $a-b$. That is to say, if the difference of the same powers of two quantities is divisible by their difference, the difference of the powers of the *next higher degree* is also divisible by that difference. Therefore, since a^4-b^4 is divisible by $a-b$, a^5-b^5 must be divisible by $a-b$; also, a^6-b^6, and so on.

The quotients obtained by dividing the difference of the powers of two quantities by the difference of those quantities, follow a simple law. Thus,

$(a^2-b^2)\div(a-b)=a+b.$

$(a^3-b^3)\div(a-b)=a^2+ab+b^2.$

$(a^4-b^4)\div(a-b)=a^3+a^2b+ab^2+b^3.$

$(a^5-b^5)\div(a-b)=a^4+a^3b+a^2b^2+ab^3+b^4.$

&c., &c., &c.

$(a^n-b^n)\div(a-b)=a^{n-1}+a^{n-2}b+a^{n-3}b^2+\ldots+a^2b^{n-3}+ab^{n-2}+b^{n-1}.$

The exponents of a decrease by unity, while those of b increase by unity.

(77.) It may also be proved that *the difference of two even powers of the same degree is divisible by the sum of their roots.* Thus,

$(a^2-b^2)\div(a+b)=a-b.$

$(a^4-b^4)\div(a+b)=a^3-a^2b+ab^2-b^3.$

$(a^6-b^6)\div(a+b)=a^5-a^4b+a^3b^2-a^2b^3+ab^4-b^5.$

&c., &c., &c.

Also, *the sum of two odd powers of the same degree is divisible by the sum of their roots.*

Thus,

$(a^3+b^3)\div(a+b)=a^2-ab+b^2.$

$(a^5+b^5)\div(a+b)=a^4-a^3b+a^2b^2-ab^3+b^4.$

$(a^7+b^7)\div(a+b)=a^6-a^5b+a^4b^2-a^3b^3+a^2b^4-ab^5+b^6.$

&c., &c., &c.

(78.) The preceding principles will enable us to resolve various algebraic expressions into their factors.

1. Resolve a^3-b^3 into its factors.

Ans. $(a^2+ab+b^2)(a-b)$

2. Resolve a^3+b^3 into its factors.
3. Resolve a^6-b^6 into four factors.
4. Resolve a^3-8b^3 into its factors.
5. Resolve $8a^3-1$ into its factors.
6. Resolve $8a^3-8b^3$ into three factors.

7. Resolve $1+27b^3$ into its factors.

8. Resolve $8a^3+27b^3$ into its factors.

(79.) One polynomial can not be divided by another polynomial containing *a letter which is not found* in the dividend; for it is impossible that one quantity multiplied by another which contains a certain letter, should give a product *not containing* that letter.

A monomial is never divisible by a polynomial, because every polynomial multiplied by another quantity gives a product containing at least *two terms* not susceptible of reduction.

Yet a binomial may be divided by a polynomial containing *any number* of terms.

Thus, a^4-b^4 is divisible by $a^3+a^2b+ab^2+b^3$, and gives for a quotient $a-b$.

So, also, a binomial may be divided by a polynomial of a hundred terms, a thousand terms, or, indeed, any finite number.

DIVISION BY DETACHED COEFFICIENTS.

(80.) We have shown, in *Art.* 64, how multiplication may sometimes be conveniently performed by operating upon the coefficients alone. The same principle is applicable to division. Thus, take the example of *Art.* 73, to divide $a^2+2ab+b^2$ by $a+b$; we may proceed as follows:

$$\begin{array}{l|l} 1+2+1 & 1+1 \\ \underline{1+1} & \overline{1+1} \\ 1+1 & \\ 1+1 & \end{array}$$

The coefficients of the quotient are $1+1$. Moreover, $a^2 \div a = a$; and therefore a is the first term of the quotient, and b the second.

Ex. 2. Divide $x^4-3ax^3-8a^2x^2+18a^3x-8a^4$ by $x^2+2ax-2a^2$

$$\begin{array}{l|l} 1-3-8+18-8 & 1+2-2 \\ \underline{1+2-2} & \overline{1-5+4} \\ -5-6+18-8 & \\ \underline{-5-10+10} & \\ 4+8-8 & \\ 4+8-8 & \end{array}$$

The coefficients of the quotient are $1-5+4$, and it remains to supply the letters. Now $x^4 \div x^2 = x^2$; and $a^4 \div a^2 = a^2$. Hence x^2, ax, and a^2 are the literal parts of the terms, and therefore the quotient is

$$x^2-5ax+4a^2.$$

Ex. 3. Divide $6a^4-96$ by $3a-6$.

Here, as we have the fourth power of a without the lower powers, we must supply the coefficients of the absent terms, as in multiplication, with zero.

```
6+ 0+0+0-96|3-6
6-12       |2+4+8+16
-----------
   12
   12-24
   -----
      24
      24-48
      -----
         48-96
         48-96
```

But $a^4 \div a = a^3$; hence the quotient is

$$2a^3+4a^2+8a+16.$$

Ex. 4. Divide $3y^3+3xy^2-4x^2y-4x^3$ by $x+y$.

Ans. $3y^2-4x^2$.

Ex. 5. Divide $8a^5-4a^4x-2a^3x^2+a^2x^3$ by $4a^2-x^2$.

Ans. $2a^3-a^2x$.

Ex. 6. Divide $a^6+4a^5-8a^4-25a^3+35a^2+21a-28$ by a^2+5a+4.

Ans. $a^4-a^3-7a^2+14a-7$.

MISCELLANEOUS EXAMPLES.

1. Divide $3a^4-19a^3b+25a^2b^2-25ab^3$ by $3a^2-4ab+5b^2$.

Ans. a^2-5ab.

2. Divide $x^4+2x^2-4x^2y^2+16xy-15$ by $x^2-2xy+5$.

Ans. $x^2+2xy-3$.

3. Divide $a^2x^2-4a^2bx+3acx+3a^2b^2+abc-10c^2$ by $ax-3ab+5c$.

Ans. $ax-ab-2c$.

4. Divide $20a^4b^6-22a^3b^7+11a^2b^8-3ab^9$ by $4a^2b^3-2ab^4+b^5$.

Ans. $5a^2b^3-3ab^4$.

5. Divide $x^6-x^4y^2+2x^2y^4-y^6$ by $x^3-x^2y+y^3$.

Ans. $x^3+x^2y-y^3$.

6. Divide $3x^7+15x^5y^2-22ax^4-2x^4y^3-10x^2y^5+14axy^3-5ax^2y^2+7a^2x$ by x^3+5xy^2-7a.

Ans. $3x^4-2xy^3-ax$.

SECTION VI.

FRACTIONS.

(81.) When a quotient is expressed as described in *Art.* 16, by placing the divisor under the dividend with a line between them, it is called a *fraction;* the dividend is called the numerator, and the divisor the denominator of the fraction. Algebraic fractions do not differ essentially from arithmetical fractions, and the same principles are applicable to both.

The following principles form the basis of most of the operations upon fractions:

1. *In order to multiply a fraction by any number, we must multiply its numerator, or divide its denominator by that number.*

Thus, the value of the fraction $\frac{ab}{a}$ is b. If we multiply the numerator by a, we obtain $\frac{a^2b}{a}$ or ab; and if we divide the denominator of the same fraction by a, we obtain also ab; that is, the original value of the fraction b has been multiplied by a.

2. *In order to divide a fraction by any number, we must divide its numerator or multiply its denominator by that number.*

Thus, the value of the fraction $\frac{a^2b}{a}$ is ab. If we divide the numerator by a, we obtain $\frac{ab}{a}$ or b; and if we multiply the denominator of the same fraction by a, we obtain $\frac{a^2b}{a^2}$ or b; that is, the original value of the fraction ab has been divided by a.

3. *The value of a fraction is not changed if we multiply or divide both numerator and denominator by the same number.*

Thus, $$\frac{ab}{a}=\frac{abx}{ax}=\frac{abxy}{axy}=b.$$

Every quantity which is not expressed under a fractional form, is called an *entire* quantity.

An algebraic expression composed partly of an entire quantity and partly of a fraction, is called a *mixed* quantity.

(82.) The proper *sign* to be prefixed to a fraction may be determined by the rules already established for division. The sign prefixed to the numerator of a fraction affects merely the *dividend;* the sign prefixed to the denominator affects merely the *divisor;* but the sign prefixed to the dividing line of a fraction affects the *quotient.*

Thus, $\frac{ab}{a}=+b$, for + divided by + gives +.

$\frac{-ab}{a}=-b$, for − divided by + gives −.

$\frac{ab}{-a}=-b$, for + divided by − gives −.

$\frac{-ab}{-a}=+b$, for − divided by − gives +.

So, also, $-\frac{ab}{a}=-b$, for this shows that the former quotient b is to be subtracted, which is done by changing its sign.

$-\frac{-ab}{a}=+b$, because the former quotient $-b$ is to be subtracted, whence it becomes $+b$.

$-\frac{ab}{-a}=+b$, for the same reason;

and $-\frac{-ab}{-a}=-b$, also for the same reason.

Hence we have the following equivalent forms:

$$\frac{ab}{a}=\frac{-ab}{-a}=-\frac{-ab}{a}=-\frac{ab}{-a}=+b;$$

also, $$\frac{-ab}{a}=\frac{ab}{-a}=-\frac{ab}{a}=-\frac{-ab}{-a}=-b.$$

That is, of the three signs belonging to the numerator, denominator, and dividing line of a fraction, *any two may be changed from + to − or from − to +, without affecting the value of the fraction.*

In the examples of fractions here employed for illustration, both numerator and denominator have consisted of monomials. The same principles are applicable to polynomials; but it must be remarked, that by the sign of the numerator we understand the *entire* numerator as distinguished from the sign of any one of its terms taken *singly*.

Thus, $-\frac{a+b+c}{d}$ is equal to $+\frac{-a-b-c}{d}$.

When *no* sign is prefixed either to the terms of a fraction or to its dividing line, + is always to be understood.

REDUCTION OF FRACTIONS.

PROBLEM I.

(83.) *To reduce a fraction to lower terms.*

RULE.

Divide both numerator and denominator by any quantity which will divide them both without a remainder.

According to Remark 3, *Art.* 81, this will not change the *value* of the fraction.

Thus, $\frac{ax}{bx}=\frac{a}{b}$.

Also, $\frac{a^2bc}{5a^2b^2}=\frac{c}{5b}$ (dividing both numerator and denominator by a^2b.)

And $\frac{ax^2}{ax+x^2}=\frac{ax}{a+x}$.

If the numerator and denominator are both divided by their *greatest* common divisor, it is evident the fraction will be reduced to its *lowest* terms. The method of finding the greatest common divisor is considered in Section XV.; but in the following examples the greatest common divisor is easily found, by resolving the quantities into factors according to methods already indicated.

EXAMPLES.

1. Reduce $\frac{cx+x^2}{a^2c+a^2x}$ to its lowest terms.

Ans. $\frac{x}{a^2}$.

2. Reduce $\frac{14a^2-7ab}{10ac-5bc}$ to its lowest terms.

Ans. $\frac{7a}{5c}$

3. Reduce $\frac{x^2-a^2}{x^4-a^4}$ to its lowest terms.

Ans. $\frac{1}{x^2+a^2}$

4. Reduce $\frac{2x^3-16x-6}{3x^3-24x-9}$ to its lowest terms.

Ans. $\frac{2}{3}$.

5. Reduce $\frac{3x^2y+3xy^2}{3x^2+6xy+3y^2}$ to its lowest terms.

Ans. $\frac{xy}{x+y}$.

6. Reduce $\frac{a^2-b^2}{a^2-2ab+b^2}$ to its lowest terms.

7. Reduce $\frac{a^3-x^3}{a^2-2ax+x^2}$ to its lowest terms.

PROBLEM II.

(84.) *To reduce a fraction to an entire or mixed quantity.*

RULE.

Divide the numerator by the denominator for the entire part, and place the remainder, if any, over the denominator for the fractional part.

Thus, $\frac{27}{5}=27\div5=5\frac{2}{5}$.

Also, $\frac{ax+a^2}{x}=(ax+a^2)\div x=a+\frac{a^2}{x}$.

EXAMPLES.

1. Reduce $\frac{ax-x^2}{x}$ to an entire quantity.

2. Reduce $\frac{ab-2a^2}{b}$ to a mixed quantity.

3. Reduce $\frac{a^2+x^2}{a-x}$ to a mixed quantity.

Ans. $a+x+\frac{2x^2}{a-x}$.

4. Reduce $\frac{x^3-y^3}{x-y}$ to an entire quantity.

5. Reduce $\frac{10x^2-5x+3}{5x}$ to a mixed quantity.

6 Reduce $\frac{8b^3-16b+7a^2b^2}{8b}$, to a mixed quantity.

PROBLEM III.

(85.) *To reduce a mixed quantity to the form of a fraction*

RULE.

Multiply the entire part by the denominator of the fraction; to the product add the numerator with its proper sign, and place the result over the denominator.

Thus, $3\frac{2}{5}=\frac{3\times5+2}{5}=\frac{15+2}{5}=\frac{17}{5}$.

This result may be proved by the preceding Rule. For $\frac{17}{5}=17\div5=3\frac{2}{5}$.

Also, $a+\frac{b}{c}=\frac{a\times c+b}{c}=\frac{ac+b}{c}$.

EXAMPLES.

1. Reduce $x+\frac{a^2-x^2}{x}$ to the form of a fraction.

Ans. $\frac{a^2}{x}$.

2. Reduce $x+\frac{ax+x^2}{2a}$ to the form of a fraction.

Ans. $\frac{3ax+x^2}{2a}$

3. Reduce $5+\frac{2x-7}{3x}$ to the form of a fraction.

4. Reduce $1+\frac{x-a-1}{a}$ to the form of a fraction.

5. Reduce $1+2x+\frac{x-3}{5x}$ to the form of a fraction.

6. Reduce $7+\frac{3b^2-8c^2}{a^2-b^2}$ to the form of a fraction.

PROBLEM IV.

(86.) *To reduce fractions to a common denominator.*

RULE.

Multiply each numerator into all the denominators, except its own, for a new numerator, and all the denominators together for a common denominator.

EXAMPLES.

1. Reduce $\frac{a}{b}$ and $\frac{c}{d}$ to a common denominator.

Ans. $\frac{ad}{bd}$; $\frac{bc}{bd}$.

Here it will be seen that the numerator and denominator of the first fraction are both multiplied by d, and in the second fraction they are both multiplied by b. The *value* of the fractions, therefore, is not changed by this operation.

2. Reduce $\frac{a}{b}$ and $\frac{a+b}{c}$ to equivalent fractions having a common denominator.

Ans. $\frac{ac}{bc}$; $\frac{ab+b^2}{bc}$.

3. Reduce $\frac{3x}{2a}$, $\frac{2b}{3c}$, and d to fractions having a common denominator.

4. Reduce $\frac{3}{4}$, $\frac{2x}{3}$, and $a+\frac{4x}{4}$ to fractions having a common denominator.

5. Reduce $\frac{a}{2}$, $\frac{3x}{7}$, and $\frac{a+x}{a-x}$ to fractions having a common denominator.

6. Reduce $\frac{x}{3}$, $\frac{x+1}{5}$, and $\frac{1-x}{1+x}$ to fractions having a common denominator.

7. Reduce $\frac{2a}{3x^2}$ and $\frac{a+2x}{4x}$ to fractions having a common denominator.

Following the Rule, we obtain

$$\frac{8ax}{12x^3} \text{ and } \frac{3ax^2+6x^3}{12x^3},$$

which fractions have a common denominator, and are equivalent to those originally proposed. Nevertheless, it may be observed, that these fractions are not reduced to their *least* common denominator, for every term is divisible by x. The least common denominator is the least common multiple of the denominators of the proposed fractions.

A common multiple of two or more numbers is any number which they will divide without a remainder; and the *least* common multiple is the least number which they will so divide. Thus, $12x^2$ is the least common multiple of $3x^2$ and $4x$; and the above fractions reduced to their least common denominator are

$$\frac{8a}{12x^2} \text{ and } \frac{3ax+6x^2}{12x^2}.$$

The least common multiple of two numbers is their product divided by their greatest common divisor.

8. Reduce $\frac{3}{14}$ and $\frac{5}{21}$ to equivalent fractions having the least common denominator.

The product of the denominators is 294, which, divided by 7 (their greatest common divisor), gives 42, the least common denominator, and the required fractions are

$$\frac{9}{42} \text{ and } \frac{10}{42}.$$

9. Reduce the fractions $\frac{7}{10}$ and $\frac{11}{15}$ to others which have the least common denominator.

10. Reduce $\frac{2a}{3bc}$ and $\frac{bd}{6b^2c^2}$ to equivalent fractions having the least common denominator.

Ans. $\frac{4ac}{6bc^2}$ and $\frac{d}{6bc^2}$.

11. Reduce $\frac{a+b}{a-b}$ and $\frac{c+d}{a^2-b^2}$ to equivalent fractions having the least common denominator.

Ans. $\frac{(a+b)^2}{a^2-b^2}$ and $\frac{c+d}{a^2-b^2}$.

PROBLEM V.

(87.) *To add fractional quantities together.*

RULE.

Reduce the fractions to a common denominator; add the numerators together, and place their sum over the common denominator.

The fractions must first be reduced to a common denominator to render them like parts of unity. Before this reduction, they must be considered as unlike quantities.

EXAMPLES.

1. What is the sum of $\frac{x}{2}$ and $\frac{x}{3}$?

Reducing to a common denominator, the fractions become

$$\frac{3x}{6} \text{ and } \frac{2x}{6}.$$

Adding the numerators, we obtain $\frac{5x}{6}$.

It is plain that three sixths of x and two sixths of x make five sixths of x.

2. Required the sum of $\frac{a}{b}$, $\frac{c}{d}$, and $\frac{e}{f}$.

Ans. $\frac{adf+bcf+bde}{bdf}$.

3. Required the sum of $\frac{a}{a+b}$ and $\frac{b}{a-b}$.

4. Required the sum of $5x$, $\frac{2a}{3x^2}$, and $\frac{a+2x}{4x}$.

5. Required the sum of $2a$, $3a+\frac{2x}{5}$, and $a+\frac{8x}{9}$.

Ans. $6a+\frac{58x}{45}$.

6. Required the sum of $a+x$, $\frac{a}{a-x}$, and $\frac{a-x}{a}$.

Ans. $a+x+2+\frac{x^2}{a^2-ax}$.

7. Required the sum of $\frac{a+b}{2}$ and $\frac{a-b}{2}$.

Ans. a.

8. Required the sum of $\frac{a}{2}$, $\frac{a-2m}{4}$, and $\frac{a+2m}{4}$.

9. Required the sum of $\frac{ma-b}{m+n}$ and $\frac{na+b}{m+n}$.

PROBLEM VI.

(88.) *To subtract one fractional quantity from another.*

RULE.

Reduce the fractions to a common denominator, subtract one numerator from the other, and place their difference over the common denominator.

EXAMPLES.

1. From $\frac{2x}{3}$ subtract $\frac{3x}{5}$.

Reducing to a common denominator, the fractions become

$$\frac{10x}{15} \text{ and } \frac{9x}{15}.$$

Hence $$\frac{10x}{15}-\frac{9x}{15}=\frac{x}{15};$$

and it is plain that *ten* fifteenths of x, diminished by *nine* fifteenths of x, equals *one* fifteenth of x.

2. From $\frac{12x}{7}$ subtract $\frac{3x}{5}$.

3. From $\frac{9x-4y}{7}$ subtract $-\frac{5x-3y}{3}$.

It must be remembered, that the minus sign before the dividing line of a fraction affects the quotient (*Art.* 82); and since a quantity is subtracted by changing its sign, the result of the subtraction in this case is

$$\frac{9x-4y}{7}+\frac{5x-3y}{3};$$

which fractions may be reduced to a common denominator, and the like terms united, as in addition.

4. From $\frac{ax}{b-c}$ subtract $\frac{ax}{b+c}$.

Ans. $\frac{2acx}{b^2-c^2}$.

5. From $2x+\frac{2+7x}{8}$ subtract $x-\frac{5x-6}{21}$.

Ans. $\frac{355x-6}{168}$

6. From $3x+\frac{x}{2b}$ subtract $x-\frac{x-a}{c}$.

7. From $\frac{a+b}{2}$ subtract $\frac{a-b}{2}$.

8. From $\frac{13a-5b}{4}$ subtract $\frac{7a-2b}{6}$.

Ans. $\frac{25a-11b}{12}$.

PROBLEM VII.

(89.) *To multiply fractional quantities together.*

RULE.

Multiply all the numerators together for a new numerator, and all the denominators together for a new denominator.

Let it be required to multiply $\frac{a}{b}$ by $\frac{c}{d}$.

First, let us multiply $\frac{a}{b}$ by c. According to Remark first of *Art.* 81, the product must be $\frac{ac}{b}$.

But the proposed multiplier was $\frac{c}{d}$; that is, we have used a multiplier d times too great. We must therefore divide the result $\frac{ac}{b}$ by d; and, according to Remark second of *Art.* 81, we obtain

$$\frac{ac}{bd},$$

which result conforms to the Rule above given.

EXAMPLES.

1. Multiply $\frac{x}{6}$ by $\frac{2x}{9}$.

Ans. $\frac{x^2}{27}$

2. What is the continued product of $\frac{x}{2}$, $\frac{4x}{5}$, and $\frac{10x}{21}$?

3. Multiply $\frac{x}{a}$ by $\frac{x+a}{a+c}$.

4. What is the continued product of $\frac{2x}{a}$, $\frac{3ab}{c}$, and $\frac{3ac}{2b}$?

Ans. $9ax$.

5. Multiply $b+\frac{bx}{a}$ by $\frac{a}{x}$.

6. Multiply $\frac{x^2-b^2}{bc}$ by $\frac{x^2+b^2}{b+c}$.

Ans. $\frac{x^4-b^4}{b^2c+bc^2}$.

7. What is the continued product of x, $\frac{x+1}{a}$, and $\frac{x-1}{a+b}$?

Ans. $\frac{x^3-x}{a^2+ab}$.

8. Multiply $\frac{a^2+b^2}{a^2-b^2}$ by $\frac{a-b}{a+b}$.

Ans. $\frac{a^2+b^2}{(a+b)^2}$.

(90.) *Ex.* 1. Multiply $\frac{1}{a^3}$ by $\frac{1}{a^2}$.

According to the preceding Article, the result must be $\frac{1}{a^5}$.

But, according to *Art.* 68, $\frac{1}{a^3}$ may be written a^{-3}; $\frac{1}{a^2}$ may be written a^{-2}; and $\frac{1}{a^5}$ may be written a^{-5}.

Therefore, $a^{-3}\times a^{-2}=a^{-5}$.

That is, *the Rule of Art.* 50 *is general, and applies to negative as well as positive exponents.*

Ex. 2. Multiply $-b^{-2}$ by b^{-3}.

Ans. $-b^{-5}$.

3. Multiply a^{-2} by a^3.
4. Multiply b^{-3} by b^3.
5. Multiply a^{-n} by a^{-m}.
6. Multiply b^{-n} by b^m.
7. Multiply $(a-b)^6$ by $(a-b)^{-3}$.

PROBLEM VIII.

(91.) *To divide one fractional quantity by another.*

RULE.

Invert the divisor, and proceed as in multiplication.

If the two fractions have the same denominator, then the quotient of the fractions will be the same as the quotient of their numerators.

Thus it is plain that $\frac{3}{12}$ is contained in $\frac{9}{12}$ as often as 3 is contained in 9.

But when the two fractions have not the same denominator, we must reduce them to this form by Problem IV.

Let it be required to divide $\frac{a}{b}$ by $\frac{c}{d}$.

Reducing to a common denominator, we have $\frac{ad}{bd}$ to be divided by $\frac{bc}{bd}$.

It is now plain that the quotient must be represented by the division of ad by bc, which gives

$$\frac{ad}{bc}$$

the same result as obtained by the above Rule.

Thus, $$\frac{a}{b} \div \frac{c}{d} = \frac{a}{b} \times \frac{d}{c} = \frac{ad}{bc}.$$

EXAMPLES.

1. Divide $\frac{x}{3}$ by $\frac{2x}{9}$.

Ans. $1\frac{1}{2}$.

2. Divide $\frac{2a}{b}$ by $\frac{4c}{d}$.

3. Divide $\frac{2x^2}{a^3+x^3}$ by $\frac{x}{x+a}$.

4. Divide $\frac{x+1}{6}$ by $\frac{2x}{3}$.

5. Divide $\frac{x-b}{8cd}$ by $\frac{3cx}{4d}$.

Ans. $\frac{x-b}{6c^2x}$.

6. Divide $\frac{2ax+x^2}{c^3-x^3}$ by $\frac{x}{c-x}$.

Ans. $\frac{2a+x}{c^2+cx+x^2}$.

7. Divide $\frac{a}{a+b}+\frac{b}{a-b}$ by $\frac{a}{a-b}-\frac{b}{a+b}$.

Ans. Unity.

(92.) *Ex.* 1. Divide $\frac{1}{a^5}$ by $\frac{1}{a^3}$.

According to the Rule of the preceding Article, we have

$$\frac{1}{a^5} \times \frac{a^3}{1} = \frac{a^3}{a^5} = \frac{1}{a^2}.$$

But $\frac{1}{a^5}$ may be written a^{-5}; $\frac{1}{a^3}$ may be written a^{-3}; and $\frac{1}{a}$ is equal to a^{-2}.

Hence $$a^{-5} \div a^{-3} = a^{-2}.$$

That is, *the Rule of Art.* 66 *is general, and applies to negative as well as positive exponents.*

Ex. 2. Divide $-b^{-5}$ by $-b^{-2}$.

Ans. b^{-3}.

3. Divide a^2 by a^{-1}.
4. Divide 1 by a^{-4}.
5. Divide $6a^n$ by $-2a^{-3}$.
6. Divide b^{m-n} by b^m.
7. Divide $12x^{-2}y^{-4}$ by $-4xy^2$.
8. Divide $(x-y)^{-4}$ by $(x-y)^{-6}$.

(93.) According to the definition, *Art.* 33, the reciprocal of a quantity is the quotient arising from dividing a unit by that quantity.

Hence the reciprocal of $\frac{a}{b}$

$$\text{is } 1 \div \frac{a}{b} = 1 \times \frac{b}{a} = \frac{b}{a}.$$

That is, *the reciprocal of a fraction is the fraction inverted.*

Thus the reciprocal of $\frac{a}{b+x}$ is $\frac{b+x}{a}$.

The reciprocal of $\frac{1}{b+c}$ is $b+c$.

Hence, *to divide by any quantity is the same as to multiply by its reciprocal, and to multiply by any quantity is the same as to divide by its reciprocal.*

(94.) The numerator or denominator of a fraction may be itself a fraction;

As $$\frac{\frac{a}{b}}{c} \text{ or } \frac{\frac{a}{b}}{\frac{c}{d}}.$$

Such expressions are easily reduced by applying the preceding principles.

Thus, $\dfrac{\left(\dfrac{a}{b}\right)}{c}$ means $\dfrac{a}{b} \div c$,

which, according to Remark second, *Art.* 81, equals $\dfrac{a}{bc}$.

Again, $\dfrac{a}{\left(\dfrac{b}{c}\right)}$ means $a \div \dfrac{b}{c}$,

which, according to *Art.* 91, equals $\dfrac{ac}{b}$.

Also, $\dfrac{\left(\dfrac{a}{b}\right)}{\left(\dfrac{c}{d}\right)}$ means the same as $\dfrac{a}{b} \div \dfrac{c}{d}$,

which, according to *Art.* 91, equals $\dfrac{ad}{bc}$.

Ex. 1. Find the value of the fraction $\dfrac{\frac{2}{3}}{\frac{4}{5}}$.

Ex. 2. Find the value of the fraction $\dfrac{2\frac{1}{2}}{\frac{3}{4}}$.

MISCELLANEOUS EXAMPLES.

1. Divide $7a^2-3x+\dfrac{m}{n}$ by $b^2-\dfrac{a}{3}$. *Ans.* $\dfrac{21a^2n-9nx+3m}{3b^2n-an}$.

2. Divide $5a^2+\dfrac{x}{4}$ by $m^2-\dfrac{x-y}{3}$. *Ans.* $\dfrac{60a^2+3x}{12m^2-4x+4y}$.

3. Divide $15x^2-\dfrac{3ab}{5c}$ by $x-\dfrac{a-b}{c}$. *Ans.* $\dfrac{75cx^2-3ab}{5cx-5a+5b}$.

4. Divide $\dfrac{2x^2-3y^2}{x}$ by $\dfrac{3x^2+2y^2}{y}$. *Ans.* $\dfrac{2x^2y-3y^3}{3x^3+2xy^2}$.

5. Divide $\dfrac{7a^3x^5}{2by^4}$ by $\dfrac{a^3+y^2}{7bx^3}$. *Ans.* $\dfrac{49a^3x^8}{2a^3y^4+2y^6}$.

6. Divide $x^2+\dfrac{x^3}{a-b}$ by $\dfrac{ab}{a-b}-y$. *Ans.* $\dfrac{ax^2-bx^2+x^3}{ab-ay+by}$.

SECTION VII.

SIMPLE EQUATIONS.

(95.) *An equation is a proposition which declares the equality of two quantities expressed algebraically.*

Thus, $x-4=b-x$, is a proposition expressing the equality of the quantities $x-4$ and $b-x$.

The quantity on the left side of the sign of equality is called the *first member* of the equation; the quantity on the right, the *second member*.

Equations are usually composed of certain quantities which are *known*, and others which are *unknown*. The known quantities are represented either by numbers or by the first letters of the alphabet, a, b, c, &c.; the unknown quantities by the last letters, x, y, z, &c.

An *identical equation* is one in which the two members are identical, or may be reduced to identity by performing the operations which are indicated in them.

Thus,
$$2x-5=2x-5$$
$$3x+4x=7x$$
$$(x+y)\,(x-y)=x^2-y^2.$$

A *root* of an equation is the value of the unknown quantity in the equation.

(96.) Equations are divided into *degrees*, according to the highest power of the unknown quantity which they contain.

Those which contain only the *first* power of the unknown quantity are called *simple equations*, or equations of *the first degree*.

As
$$ax+b=cx+d.$$

Those in which the highest power of the unknown quantity

is a *square*, are called *quadratic equations*, or equations of *the second degree.*

As $$4x^2-2x=5-x^2.$$

Those in which the highest power is a *cube*, are called *cubic equations*, or equations of *the third degree.*

As $$x^3+px^2=2q.$$

So, also, we have *biquadratic equations*, or equations of *the fourth degree;* equations of the *fifth, sixth,* — — — — — *n*th degree.

Thus, $x^n+px^{n-1}=r$, is an equation of the *n*th degree.

In general, *the degree of an equation is determined by the highest of the exponents with which the unknown quantity is affected.*

(97.) *Numerical equations* are those which contain only particular numbers, with the exception of the unknown quantity, which is always denoted by a letter.

Thus, $x^3+4x^2=3x+12$ is a numerical equation.

Literal equations are those in which the known quantities are represented by letters, or by letters and numbers.

Thus, $$\left.\begin{array}{r} x^3+px^2+qx=r \\ x^4-3px^3+5qx^2=5 \end{array}\right\} \text{are literal equations.}$$

To *solve* an equation is to *find the value of the unknown quantity*, or to find a number which, substituted for the unknown quantity in the equation, renders the first member identical with the second.

The difficulty of solving equations depends upon their *degree*, and *the number of unknown quantities.* We will begin with the most simple case.

SIMPLE EQUATIONS CONTAINING BUT ONE UNKNOWN QUANTITY.

(98.) The various operations which we perform upon equations in order to deduce the value of the unknown quantities, are founded upon the following principles:

1. If to two equal quantities the same quantity be *added*, the sums will be equal.

2. If from two equal quantities the same quantity be *subtracted*, the remainders will be equal.

3. If two equal quantities be *multiplied* by the same quantity, the products will be equal.

4. If two equal quantities be *divided* by the same quantity, the quotients will be equal.

(99.) The unknown quantity may be combined with the known quantities in the given equation by the operations of *addition, subtraction, multiplication,* or *division.*

We shall consider these different cases in succession.

I. The unknown quantity may be combined with known quantities by *addition.*

Let it be required to solve the equation

$$x+6=24.$$

If from the two equal quantities, $x+6$ and 24, we subtract the same quantity 6, the remainders will be equal, according to the last Article, and we shall have

$$x+6-6=24-6,$$
$$\text{or } x=24-6,$$
$$=18, \text{ the value of } x \text{ required}$$

So, also, in the equation

$$x+a=b,$$

subtracting a from each of the equal quantities, $x+a$ and b, the result is

$$x=b-a, \text{ the value of } x \text{ required.}$$

(100.) II. The unknown quantity may be combined with known quantities by *subtraction.*

Let the equation be

$$x-6=24.$$

If to the two equal quantities, $x-6$ and 24, the same quantity 6 be added, the sums will be equal, according to *Art.* 98 and we have

$$x-6+6=24+6,$$
$$\text{or } x=30, \text{ the value of } x \text{ required.}$$

So, also, in the equation

$$x-a=b,$$

adding a to each of these equal quantities, the result is

$$x=b+a, \text{ the value of } x \text{ required.}$$

From the preceding examples, it follows that

We may transpose any term of an equation from one member to the other by changing its sign.

We may change the sign of every term of an equation without destroying the equality.

This is, in fact, the same thing as transposing *every term* in each member of the equation.

If the same quantity appear in each member of the equation affected with the same sign, it may be suppressed.

(101.) III. The unknown quantity may be combined with known quantities by *multiplication.*

Let the equation be

$$6x=24.$$

If we divide each of the equal quantities, $6x$ and 24, by the same quantity 6, the quotients will be equal, and we shall have

$$x=\frac{24}{6},$$

$$=4, \text{ the value of } x \text{ required.}$$

So, also, in the equation

$$ax=b,$$

dividing each of these equals by a, the result is

$$x=\frac{b}{a}, \text{ the value of } x \text{ required.}$$

From this it follows, that

When the unknown quantity is multiplied by a known quantity, the equation is solved by dividing both members by this known quantity.

(102.) IV. The unknown quantity may be combined with known quantities by *division.*

Let the equation be

$$\frac{x}{6}=24.$$

If we multiply each of the equal quantities, $\frac{x}{6}$ and 24, by the same quantity 6, the products will be equal, and we shall have

$$x=144, \text{ the value of } x \text{ required.}$$

So, also, in the equation

$$\frac{x}{a}=b.$$

multiplying each of these equals by a, the result is

$$x=ab, \text{ the value of } x \text{ required.}$$

From this it follows, that

When the unknown quantity is divided by a known quantity,

the equation is solved by multiplying both members by this known quantity.

(103.) V. Several terms of an equation may be *fractional.*

Let the equation be

$$\frac{x}{2}=\frac{2}{3}+\frac{4}{5}.$$

Multiplying each of these equals by 2, the result is

$$x=\frac{4}{3}+\frac{8}{5}.$$

Multiplying each of these last equals by 3, we obtain

$$3x=4+\frac{24}{5};$$

and multiplying again by 5, we obtain

$$15x=20+24,$$

an equation free from fractions.

We might have obtained the same result by multiplying the original equation at once by the product of all the denominators.

Thus, multiplying by $2\times3\times5$, we have

$$\frac{30x}{2}=\frac{60}{3}+\frac{120}{5},$$

or reducing, we have

$$15x=20+24, \text{ as before.}$$

So, also, in the equation

$$\frac{x}{a}=\frac{b}{c}+\frac{d}{e},$$

multiplying successively by all the denominators, or by $a\ c\ e$ at once, we obtain

$$\frac{acex}{a}=\frac{abce}{c}+\frac{acde}{e}.$$

Canceling from each term the letter which is common to its numerator and denominator, we have

$$cex=abe+acd,$$

an equation clear of fractions.

Hence it appears that

An equation may be cleared of fractions by multiplying each member into all the denominators.

(104.) From the preceding remarks, we deduce the following

RULE FOR THE SOLUTION OF A SIMPLE EQUATION CONTAINING ONE UNKNOWN QUANTITY.

1. *Clear the equation of fractions, and perform in both members all the algebraic operations indicated.*

2. *Transpose all the terms containing the unknown quantity to one side, and all the remaining terms to the other side of the equation, and reduce each member to its most simple form.*

3. *Divide each member by the coefficient of the unknown quantity.*

EXAMPLES.

1. Given $5x+8=4x+10$, to find the value of x.

Transposing $4x$ to the first member of the equation, and 8 to the second member, taking care to change their signs (*Art.* 100), we have

$$5x-4x=10-8.$$

Uniting similar terms, $x=2$.

In order to verify this result, put 2 in the place of x wherever it occurs in the original equation, and we shall obtain

$$5\times2+8=4\times2+10.$$

That is, $10+8=8+10$,

or $18=18$,

an identical equation, which proves that we have found the correct value of x.

2. Given $x-7=\frac{x}{5}+\frac{x}{3}$, to find the value of x.

Multiplying every term of the equation by 5 and also by 3, in order to clear it of fractions (*Art.* 103), we obtain

$$15x-105=3x+5x.$$

Hence, by transposition,

$$15x-3x-5x=105,$$

or $7x=105$,

and therefore $x=\frac{105}{7}=15$.

To verify this result, put 15 in the place of x in the original equation, and we have

$$15-7=\frac{15}{5}+\frac{15}{3}.$$

That is, $15-7=3+5$,

or $8=8$,

an identical equation.

3. Given $3ax-4ab=2ax-6ac$, to find the value of x in terms of b and c.

Dividing every term by a, we have

$$3x-4b=2x-6c.$$

By transposition,

$$3x-2x=4b-6c,$$

or $x=4b-6c$.

This result may be verified in the same manner as the preceding.

4. Given $3x^2-10x=8x+x^2$, to find the value of x.

Ans. $x=9$.

5. Given $\frac{a(d^2+x^2)}{dx}=ac+\frac{ax}{d}$, to find x.

Ans. $x=\frac{d}{c}$.

6. Given $\frac{x-5}{4}+6x=\frac{284-x}{5}$, to find x.

Ans. $x=9$.

7. Given $\frac{ab}{x}=bc+d+\frac{1}{x}$, to find x.

Ans. $x=\frac{ab-1}{bc+d}$.

8. Given $3x+\frac{2x+6}{5}=5+\frac{11x-37}{2}$, to find x.

Ans. $x=7$.

9. Given $5ax-2b+4bx=2x+5c$, to find x.

Ans. $x=\frac{5c+2b}{5a+4b-2}$.

10. Given $x+\frac{3x-5}{2}=12-\frac{2x-4}{3}$, to find the value of x.

Ans. $x=5$.

11. Given $21+\frac{3x-11}{16}=\frac{5x-5}{8}+\frac{97-7x}{2}$, to find x.

(105.) An equation may always be cleared of fractions by multiplying each member into *all* the denominators according to *Art.* 103. But sometimes the same object may be attained by a less amount of multiplication.

Thus, in the preceding example, the equation may be cleared of fractions by multiplying each term by 16, instead of $16\times8\times2$, and it is important to avoid all useless multiplication. In general, it is sufficient to multiply by the *least common multiple* of all the denominators. See *Art.* 86.

12. Given $3x-\frac{x-4}{4}-4=\frac{5x+14}{3}-\frac{1}{12}$, to find x.

13. Given $3x-a+cx=\frac{a+x}{3}-\frac{b-x}{a}$, to find x.

Ans. $x=\frac{4a^2-3b}{8a+3ac-3}$.

14. Given $\frac{3x}{a}-c+\frac{x}{b}=4x+\frac{2x}{d}$, to find x.

Ans. $x=\frac{abcd}{3bd+ad-4abd-2ab}$.

15. Given $(a+x)(b+x)-a(b+c)=\frac{a^2c}{b}+x^2$, to find x.

Ans. $x=\frac{ac}{b}$.

16. Given $\frac{17-3x}{5}-\frac{4x+2}{3}=5-6x+\frac{7x+14}{3}$, to find x.

17. Given $x-\frac{3x-3}{5}+4=\frac{20-x}{2}-\frac{6x-8}{7}+\frac{4x-4}{5}$, to find x.

18. Given $\frac{7x+16}{21}-\frac{x+8}{4x-11}=\frac{x}{3}$, to find x.

19. Given $\frac{6x+7}{9}+\frac{7x-13}{6x+3}=\frac{2x+4}{3}$, to find x.

20. Given $\frac{5}{6}ab+\frac{4}{5}ac-\frac{2}{3}cx=\frac{3}{4}ac+2ab-6cx$, to find the value of x.

Ans. $x=\frac{70ab-3ac}{320c}$.

SOLUTION OF PROBLEMS.

(106.) The solution of a Problem by Algebra consists of two distinct parts:

1. To express the conditions of the problem algebraically; that is, to *form* the equation.

2. To *solve* the equation.

The second operation has already been explained, but the first is often more embarrassing to beginners than the second. Sometimes the statement of a problem furnishes the equation directly; and sometimes it is necessary to deduce from the statement new conditions, which are to be expressed algebraically. The former are called *explicit* conditions; and those which are deduced from them, *implicit* conditions.

It is impossible to give a general rule which will enable us to translate every problem into algebraic language. The power of doing this with facility can only be acquired by reflection and practice.

The following directions may be found of some service.

Denote one of the required quantities by x; *then, by means of this letter, with the algebraic signs, perform the same operations which would be necessary to verify its value if it was already known.*

Problem 1. What number is that, to the double of which if 16 be added, the sum is equal to four times the required number?

Let x represent the number required.

The double of this will be $2x$.

This increased by 16 should equal $4x$.

Hence, by the conditions, $2x+16=4x$.

The problem is now translated into algebraic language, and it only remains to solve the equation in the usual way.

Transposing, we obtain

$$16=4x-2x=2x,$$

and $$8=x,$$

or $$x=8.$$

To verify this number, we have but to double 8, and add 16 to the result; the sum is 32, which is equal to four times 8, according to the conditions of the problem.

Prob. 2. What number is that, the double of which exceeds its half by 6?

Let $x =$ the number required.

Then, by the conditions,

$$2x-\frac{x}{2}=6.$$

Clearing of fractions,

$$4x-x=12,$$

or $$3x=12.$$

Hence $$x=4.$$

To verify this result, double 4, which makes 8, and diminish it by the half of 4, or 2; the result is 6, according to the conditions of the problem.

Prob. 3. The sum of two numbers is 8, and their difference 2. What are those numbers?

Let x = the least number.

Then $x+2$ will be the greater number.

The sum of these is $2x+2$, which is required to equal 8. Hence we have

$$2x+2=8.$$

By transposition, $2x=8-2=6,$

and $x=3$, the least number.

Also, $x+2=5$, the greater number.

Verification. $\left.\begin{array}{l}5+3=8\\5-3=2\end{array}\right\}$ according to the conditions.

The following is a generalization of the preceding Problem.

Prob. 4. The sum of two numbers is a, and their difference b. What are those numbers?

Let x represent the least number.

Then $x+b$ will represent the greater number.

The sum of these is $2x+b$, which is required to equal a.

Hence we have

$$2x+b=a.$$

By transposition, $2x=a-b,$

or $x=\frac{a-b}{2}=\frac{a}{2}-\frac{b}{2}$, the less number.

Hence $x+b=\frac{a}{2}-\frac{b}{2}+b=\frac{a}{2}+\frac{b}{2}$, the greater number.

As these results are independent of any particular value attributed to the letters a and b, it follows that

Half the difference of two quantities, added to half their sum, is equal to the greater; and

Half the difference subtracted from half the sum is equal to the less.

The expressions $\frac{a}{2}+\frac{b}{2}$ and $\frac{a}{2}-\frac{b}{2}$ are called *formulas*, because they may be regarded as comprehending the solution of all questions of the *same kind;* that is, of all problems in which we have given the sum and difference of two quantities.

Thus, let $\left.\begin{matrix} a=8 \\ b=2 \end{matrix}\right\}$ as in the preceding problem.

Then $\frac{a}{2}+\frac{b}{2}=\frac{8+2}{2}=5$, the greater number.

And $\frac{a}{2}-\frac{b}{2}=\frac{8-2}{2}=3$, the less number.

Given the sum of two numbers,	their difference	required the numbers.
= 10;	= 6;	
12	" 2	"
23	" 11	"
100	" 50	"
100	" 1	"
5	" $\frac{1}{2}$	"
10	" $\frac{1}{2}$	"

Prob. 5. From two towns which are 54 miles distant, two travelers set out at the same time with an intention of meeting. One of them goes 4 miles and the other 5 miles per hour. In how many hours will they meet?

Let x represent the required number of hours.

Then $4x$ will represent the number of miles one traveled,

and $5x$ the number the other traveled;

and since they meet, they must together have traveled the whole distance.

Consequently, $4x+5x=54$.

Hence $9x=54$,

or $x=6$.

Proof. In 6 hours, at 4 miles an hour, one would travel 24 miles; the other, at 5 miles an hour, would travel 30 miles. The sum of 24 and 30 is 54 miles, which is the whole distance.

This Problem may be generalized as follows:

Prob. 6. From two points which are a miles apart, two bodies move toward each other, the one at the rate of m miles per hour, the other at the rate of n miles per hour. In how many hours will they meet?

Let x represent the required number of hours.

Then mx will represent the number of miles one body moves,

and nx the miles the other body moves,
and we shall obviously have

$$mx+nx=a.$$

Hence $x=\frac{a}{m+n}$.

This is a general formula, comprehending the solution of all problems of this kind. Thus,

Let the distance		one body moves		the other		miles per hour
	=150;		6;		4	
	90		8		7	
	135		15		12	
	210		20		15	

Required the time of meeting.

We see that an infinite number of problems may be proposed, all similar to Prob. 5; but they are all solved by the formula of Prob. 6. We also see what is necessary in order that the answers may be obtained in *whole numbers*. The given distance (a) must be exactly divisible by $m+n$.

Prob. 7. A gentleman meeting three poor persons, divided 60 cents among them; to the second he gave twice, and to the third three times as much as to the first. What did he give to each?

Let $x =$ the sum given to the first.

Then $2x =$ the sum given to the second,
and $3x =$ the sum given to the third.

Then, by the conditions,

$$x+2x+3x=60.$$

That is, $6x=60$,

or $x=10$.

Therefore he gave 10, 20, and 30 cents to them respectively. The learner should verify this, and all the subsequent results.

The same problem generalized.

Prob. 8. Divide the number a into three such parts, that the second may be m times, and the third n times as great as the first.

$$Ans.\ \frac{a}{1+m+n};\ \frac{ma}{1+m+n};\ \frac{na}{1+m+n}.$$

What is necessary in order that the preceding values may be expressed in whole numbers?

Prob. 9. A bookseller sold 10 books at a certain price, and afterward 15 more at the same rate. Now at the last sale he received 25 dollars more than at the first. What did he receive for each book?

Ans. Five dollars.

The same Problem generalized.

Prob. 10. Find a number such that when multiplied successively by m and by n, the difference of the products shall be a

$$Ans.\ \frac{a}{m-n}.$$

Prob. 11. A gentleman dying, bequeathed 1000 dollars to three servants. A was to have twice as much as B, and B three times as much as C. What were their respective shares?

Ans. A received \$600, B \$300, and C \$100.

Prob. 12. Divide the number a into three such parts that the second may be m times as great as the first, and the third n times as great as the second.

$$Ans.\ \frac{a}{1+m+mn};\ \frac{ma}{1+m+mn};\ \frac{mna}{1+m+mn}.$$

Prob. 13. A hogshead which held 120 gallons was filled with a mixture of brandy, wine, and water. There were 10 gallons of wine more than there were of brandy, and as much water as both wine and brandy. What quantity was there of each?

Ans. Brandy 25 gallons, wine 35, and water 60 gallons.

Prob. 14. Divide the number a into three such parts, that the second shall exceed the first by m, and the third shall be equal to the sum of the first and second.

$$Ans.\ \frac{a-2m}{4};\ \frac{a+2m}{4};\ \frac{a}{2}.$$

Prob. 15. A person employed four workmen, to the first of whom he gave 2 shillings more than to the second; to the second 3 shillings more than to the third; and to the third 4 shillings more than to the fourth. Their wages amount to 32 shillings. What did each receive?

Ans. They received 12, 10, 7, and 3 shillings respectively.

Prob. 16. Divide the number a into four such parts, that the second shall exceed the first by m, the third shall exceed the second by n, and the fourth shall exceed the third by p.

$$\textit{Ans.}\ \frac{a-3m-2n-p}{4};\ \frac{a+m-2n-p}{4};$$
$$\frac{a+m+2n-p}{4};\ \frac{a+m+2n+3p}{4}.$$

(107.) Problems which involve several unknown quantities may often be solved by the use of a single unknown letter. Most of the preceding examples are of this kind. In general, when we have given the sum or difference of two quantities, both of them may be expressed by means of the same letter. For the difference of two quantities added to the less must be equal to the greater; and if one of two quantities be subtracted from their sum, the remainder will be equal to the other.

Prob. 17. At a certain election 36000 votes were polled; and the candidate chosen wanted but 3000 of having twice as many votes as his opponent. How many voted for each?

Let $x=$ the number of votes for the unsuccessful candidate
Then $36000-x=$ the number the successful one had,
And $36000-x+3000=2x$.

Ans. 13000 and 23000.

Prob. 18. Divide the number a into two such parts, that one part increased by b shall be equal to m times the other part.

$$\textit{Ans.}\ \frac{ma-b}{m+1};\ \frac{a+b}{m+1}.$$

Prob. 19. A train of cars moving at the rate of 20 miles per hour, had been gone three hours, when a second train followed at the rate of 25 miles per hour. In what time will the second train overtake the first?

Let $x=$ the number of hours the second train is in motion,
$x+3=$ the time of the first train.
Then $25x=$ the number of miles traveled by the second train,
$20(x+3)=$ the miles traveled by the first train.

But at the time of meeting they must both have traveled the same distance.

Therefore $25x=20x+60$.

By transposition, $5x=60$,

and $x=12$.

Proof. In 12 hours, at 25 miles per hour, the second train goes 300 miles; and in 15 hours, at 20 miles per hour, the first train also goes 300 miles; that is, it is overtaken by the second train.

Prob. 20. Two bodies move in the same direction from two places at a distance of a miles apart; the one at the rate of n miles per hour, the other pursuing at the rate of m miles per hour. When will they meet?

Ans. In $\frac{a}{m-n}$ hours.

This Problem, it will be seen, is essentially the same as Prob. 10.

Prob. 21. Divide the number 197 into two such parts, that four times the greater may exceed five times the less by 50.

Ans. 82 and 115.

Prob. 22. Divide the number a into two such parts, that m times the greater may exceed n times the less by b.

Ans. $\frac{ma-b}{m+n}$; $\frac{na+b}{m+n}$.

When $n=1$, this Problem reduces to Problem 18.
When $b=0$, this Problem reduces to Problem 24.

Prob. 23. A prize of 2329 dollars was divided between two persons, A and B, whose shares were in the ratio of 5 to 12. What was the share of each?

Beginners almost invariably put x to represent one of the quantities sought in a problem; but a solution may often be very much simplified by pursuing a different method. Thus, in the preceding problem, we may put x to represent one fifth of A's share. Then $5x$ will be A's share, and $12x$ will be B's, and we shall have the equation

$$5x+12x=2329,$$

and $x=137$,

consequently their shares were 685 and 1644 dollars.

Prob. 24. Divide the number a into two such parts, that the first part may be to the second as m to n.

Ans. $\frac{ma}{m+n}$; $\frac{na}{m+n}$.

Prob. 25. What number is that whose third part exceeds its fourth part by 16?

Let $12x=$ the number.

Then $4x-3x=16$,

or $x=16$.

Therefore the number $=12\times16=192$.

Prob. 26. Find a number such that when it is divided successively by m and by n, the difference of the quotients shall be a.

Ans. $\frac{mna}{n-m}$.

Prob. 27. What two numbers are as 2 to 3, to each of which, if four be added, the sums will be as 5 to 7?

A strict adherence to system would have required this example to be placed after the subject of Proportion, which is treated of in Section XIII. It is, however, only necessary to assume one simple principle which is employed in Arithmetic, viz., If four quantities are proportional, *the product of the extremes is equal to the product of the means.*

Thus, if $a:b::c:d$.

Then $ad=bc$.

In the preceding Problem, let $2x$ and $3x$ be the numbers.

Then $2x+4:3x+4::5:7$,

and by the last principle,

$$14x+28=15x+20.$$

Prob. 28. What two numbers are as m to n, to each of which, if a be added, the sums shall be as p to q?

Ans. $\frac{ma(p-q)}{mq-np}$; $\frac{na(p-q)}{mq-np}$.

Prob. 29. A gentleman divides a dollar among 12 children, giving to some 9 cents each, and to the rest 7 cents. How many were there of each class?

Prob. 30. Divide the number a into two such parts, that if the first is multiplied by m and the second by n, the sum of the products shall be b.

$$Ans.\ \frac{b-na}{m-n};\ \frac{ma-b}{m-n}.$$

Prob. 31. If the sun moves every day one degree, and the moon thirteen, and the sun is now 60 degrees in advance of the moon, when will they be in conjunction for the first time, second time, and so on?

Prob. 32. If two bodies move in the same direction upon the circumference of a circle which measures a miles, the one at the rate of n miles per day, the other pursuing at the rate of m miles per day, when will they meet for the first time, second time, &c., supposing them to be b miles apart at starting?

$$Ans.\ \text{In } \frac{b}{m-n};\ \frac{a+b}{m-n};\ \frac{2a+b}{m-n},\ \text{\&c., days.}$$

It will be seen that this Problem includes Prob. 20.

Prob. 33. Divide the number 12 into two such parts, that the difference of their squares may be 48.

Prob. 34. Divide the number a into two such parts, that the difference of their squares may be b.

$$Ans.\ \frac{a^2-b}{2a};\ \frac{a^2+b}{2a}.$$

Prob. 35. The estate of a bankrupt, valued at 21000 dollars, is to be divided among three creditors according to their respective claims. The debts due to A and B are as 2 to 3, while B's claims and C's are in the ratio of 4 to 5. What sum must each receive?

Prob. 36. Divide the number a into three parts, which shall be to each other as $m:n:p$.

$$Ans.\ \frac{ma}{m+n+p};\ \frac{na}{m+n+p};\ \frac{pa}{m+n+p}.$$

When $p=1$, Prob. 36 reduces to the same form as Prob. 8.

Prob. 37. A grocer has two kinds of tea, one worth 72 cents per pound, the other 40 cents. How many pounds of each must be taken to form a chest of 80 pounds, which shall be worth 60 cents?

Ans. 50 pounds at 72 cents, and 30 pounds at 40 cents

Prob. 38. A grocer has two kinds of tea, one worth a cents per pound, the other b cents. How many pounds of each must be taken to form a mixture of n pounds, which shall be worth c cents?

$$Ans.\ \frac{n(c-b)}{a-b} \text{ pounds at } a \text{ cents,}$$

$$\text{and } \frac{n(a-c)}{a-b} \text{ pounds at } b \text{ cents.}$$

Prob. 39. A can perform a piece of work in 6 days; B can perform the same work in 8 days; and C can perform the same work in 24 days. In what time will they finish it if all work together?

Prob. 40. A can perform a piece of work in a days, B in b days, and C in c days. In what time will they perform it if all work together?

$$Ans.\ \frac{abc}{ab+ac+bc} \text{ days.}$$

Prob. 41. There are three workmen, A, B, and C. A and B together can perform a piece of work in 27 days; A and C together in 36 days; and B and C together in 54 days. In what time could they finish it if all worked together?

A and B together can perform $\frac{1}{27}$ of the work in one day
A and C " $\frac{1}{36}$ " one "
B and C " $\frac{1}{54}$ " one "
Therefore, adding these three results,

$$2A+2B+2C \text{ can perform } \frac{1}{27}+\frac{1}{36}+\frac{1}{54} \text{ in one day.}$$
$$=\frac{1}{12} \text{ in one day.}$$

Therefore, A, B, and C together can perform $\frac{1}{24}$ of the work in one day; that is, they can finish it in 24 days. If we put x to represent the time in which they would all finish it, then they would together perform $\frac{1}{x}$ part of the work in one day, and we should have

$$\frac{1}{27}+\frac{1}{36}+\frac{1}{54}=\frac{2}{x}.$$

Prob. 42. A and B can perform a piece of labor in a days A and C together in b days; and B and C together in c days In what time could they finish it if all work together?

$$Ans.\ \frac{2abc}{ab+ac+bc} \text{ days.}$$

This result, it will be seen, is of the same form as that of Problem 40.

Prob. 43. A broker has two kinds of change. It takes 20 pieces of the first to make a dollar, and 4 pieces of the second to make the same. Now a person wishes to have 8 pieces for a dollar. How many of each kind must the broker give him?

Prob. 44. A has two kinds of change; there must be a pieces of the first to make a dollar, and b pieces of the second to make the same. Now B wishes to have c pieces for a dollar. How many pieces of each kind must A give him?

Ans. $\frac{a(c-b)}{a-b}$ of the first kind: $\frac{b(a-c)}{a-b}$ of the second.

Prob. 45. Divide the number 45 into four such parts, that the first increased by 2, the second diminished by 2, the third multiplied by 2, and the fourth divided by 2, shall all be equal.

In solving examples of this kind, several unknown quantities are usually introduced, but this practice is worse than superfluous. The four parts into which 45 is to be divided, may be represented thus:

The first $=x-2$,
second $=x+2$,
third $=\frac{x}{2}$,
fourth $=2x$;

for if the first expression be increased by 2, the second diminished by 2, the third multiplied by 2, and the fourth divided by 2, the result in each case will be x. The sum of the four parts is $4\frac{1}{2}x$, which must equal 45.

Hence $x=10$.

Therefore the parts are 8, 12, 5, and 20.

Prob. 46. Divide the number a into four such parts, that the first increased by m, the second diminished by m, the third multiplied by m, and the fourth divided by m, shall all be equal.

Ans. $\frac{ma}{(m+1)^2}-m$; $\frac{ma}{(m+1)^2}+m$; $\frac{a}{(m+1)^2}$; $\frac{m^2a}{(m+1)^2}$.

Prob. 47. A merchant maintained himself for three years at an expense of \$500 a year; and each year augmented that part of his stock which was not thus expended by one third tnereof. At the end of the third year his original stock was doubled. What was that stock?

Prob. 48. A merchant supported himself for three years at an expense of a dollars per year; and each year augmented that part of his stock which was not thus expended by one third thereof. At the end of the third year his original stock was doubled. What was that stock?

Ans. $\frac{148a}{10}$.

Prob. 49. A father, aged 54 years, has a son aged 9 years. In how many years will the age of the father be four times that of the son?

Prob. 50. The age of a father is represented by a, the age of his son by b. In how many years will the age of the father be n times that of the son?

Ans. $\frac{a-nb}{n-1}$.

SECTION VIII.

SIMPLE EQUATIONS CONTAINING TWO OR MORE UNKNOWN QUANTITIES.

(108.) In the examples which have been hitherto given, each problem has contained but one unknown quantity; or, if there have been more, they have been so related to each other that all have been expressed by means of the same letter. This, however, can not always be done, and we are now to consider how equations of this kind are resolved.

If we have *two* equations, with two unknown quantities, we must endeavor to deduce from them *a single equation*, containing only *one* unknown quantity. We must, therefore, make one of the unknown quantities disappear, or, as it is termed, we must *eliminate* it. There are three different methods of elimination which may be practiced.

The *first* is by substitution,
" *second* " comparison,
" *third* " addition and subtraction.

ELIMINATION BY SUBSTITUTION.

(109.) Let it be proposed to solve the system of equations

$$\left.\begin{array}{r} x+y=12 \\ x-y=\ 6. \end{array}\right\} \quad (1.)$$

From the second equation, we find the value of x in terms of y, which gives

$$x=y+6.$$

Substituting the expression $y+6$ for x in the first equation, it becomes

$$y+6+y=12;$$

from which we find that $y=3$; and since we have already seen that $x=y+6$, we find that $x=3+6=9$.

To verify these values, substitute them for x and y in the original equations, and we shall obtain

$$9+3=12$$
$$9-3=\ 6.$$

Again, take the equations

$$\left.\begin{array}{l}2x+3y=13\\5x+4y=22.\end{array}\right\}\quad(2.)$$

From the first equation we find

$$y=\frac{13-2x}{3}.$$

Substituting this value of y in the second equation, it becomes

$$5x+4\times\frac{13-2x}{3}=22,$$

an equation containing only x, which, when solved, gives

$$x=2,$$

and this value of x, substituted in either of the original equations, gives

$$y=3.$$

The method thus exemplified is expressed in the following

RULE.

Find an expression for the value of one of the unknown quantities in one of the equations; then substitute this value in the place of its equal in the other equation.

ELIMINATION BY COMPARISON.

(110.) To illustrate this method, take equations (1.) of the preceding Article. Derive from each equation an expression for x in terms of y, and we shall have

$$x=12-y,$$
$$x=\ 6+y.$$

These two values of x must be equal to each other, and by *comparing* them we shall obtain

$$12-y=6+y,$$

an equation involving only one unknown quantity;
whence $$y=3.$$

Substituting this value of y in the expression $x=6+y$, and we find $x=9$, as before.

Again, take equations (2.) of the preceding Article.

From equation first, we find

$$y=\frac{13-2x}{3},$$

ana from equation second,

$$y=\frac{22-5x}{4}.$$

Putting these values of y equal to each other, we have

$$\frac{13-2x}{3}=\frac{22-5x}{4},$$

an equation containing only x, whence we obtain

$$x=2.$$

Substituting this value of x in either of the preceding expressions for y, we find

$$y=3.$$

The method thus exemplified is expressed in the following

RULE.

Find an expression for the value of the same unknown quantity in each of the equations, and form a new equation by putting one of these values equal to the other.

ELIMINATION BY ADDITION AND SUBTRACTION.

(111.) To illustrate this method, take equations (1.) of *Art.* 109. Since the coefficients of y in the two equations are equal and have contrary signs, we may eliminate this letter by adding the two equations together, whence we obtain

$$2x=18,$$
$$\text{or } x=9.$$

We may now deduce the value of y by substituting the value of x in one of the original equations. Taking the first for example, we have

$$9+y=12,$$

whence $y=3.$

Since the coefficients of x are equal in the two original equations, we might have eliminated this letter by subtracting one equation from the other. Subtracting the first from the second, we obtain

$$2y=6,$$
$$\text{or}\quad y=3.$$

Let us apply the same method to equations (2.) of *Art.* 109. We perceive that if we could deduce from the proposed equations two other equations, in which the coefficients of y should be equal, the elimination of y might be effected by *subtracting* one of these new equations from the other.

It is easily seen that we shall obtain two equations of the form required, if we multiply all the terms of each equation by the coefficient of y in the other. Multiplying, therefore, all the terms of equation first by 4, and all the terms of equation second by 3, they become

$$8x+12y=52,$$
$$15x+12y=66.$$

Subtracting the former of these equations from the latter, we find

$$7x=14,$$
whence $$x=2.$$

In like manner, in order to eliminate x, multiply the first of the proposed equations by 5, and the second by 2, they will become

$$10x+15y=65,$$
$$10x+8y=44.$$

Subtracting the latter of these two equations from the former, we have

$$7y=21,$$
whence $$y=3.$$

This last method is expressed in the following

RULE.

Multiply or divide the equations, if necessary, in such a manner, that one of the unknown quantities shall have the same coefficient in both. Then subtract one equation from the other, if the signs of these coefficients are the same, or add them together if the signs are different.

EXAMPLES.

(112.) *Ex.* 1. Given $\left.\begin{array}{l}5x+4y=58\\3x+7y=67\end{array}\right\}$ to find the values of x and y.

By the first method.

From the second equation we find

$$3x=67-7y.$$

Therefore $$x=\frac{67-7y}{3}.$$

Substituting this value of x in the first equation,

$$5\times\frac{67-7y}{3}+4y=58.$$

Hence $$335-35y+12y=174.$$

By transposition, $$335-174=35y-12y,$$

or $$161=23y.$$

Therefore $$y=7.$$

Substituting this value of y in the expression for the value of x given above, it becomes

$$\frac{67-7\times7}{3}=\frac{67-49}{3}=\frac{18}{3}=6.$$

Thus we have $$y=7, \text{ and } x=6.$$

By the second method.

From the first equation we find

$$5x=58-4y,$$

whence $$x=\frac{58-4y}{5}.$$

From the second equation, $x=\dfrac{67-7y}{3}.$

Therefore $$\frac{58-4y}{5}=\frac{67-7y}{3}.$$

Clearing of fractions, $174-12y=335-35y.$

By transposition, $$35y-12y=335-174,$$

or $$23y=161.$$

Therefore $$y=7,$$

whence, as before, $$x=6.$$

By the third method.

Multiplying the second equation by 5 and the first by 3, we obtain

$$15x+35y=335,$$

and $$15x+12y=174.$$

By subtraction, $$23y=161,$$

or $$y=\ \ 7.$$

Whence, from equation first,

$$5x=58-4y=58-28=30,$$

and therefore $$x=6.$$

Thus the same example may be solved by either of the three methods, and each method has its advantages in particular cases. Generally, however, the first two methods give rise to fractional expressions which occasion inconvenience in practice, while the third method is not liable to this objection. When the coefficient of one of the unknown quantities in one of the equations is equal to unity, this inconvenience does not occur, and the method by substitution may be preferable; the third will, however, commonly be found most convenient.

Ex. 2. Given $\left.\begin{aligned}11x+3y&=100\\ 4x-7y&=\ \ 4\end{aligned}\right\}$ to find the values of x and y.

Multiplying the first equation by 7 and the second by 3, we obtain

$$77x+21y=700,$$

$$12x-21y=\ \ 12.$$

Therefore, by addition, $$89x=712,$$

or $$x=\ \ 8.$$

From equation first, $3y=100-11x,$

$$=100-88=12,$$

and $$y=4.$$

These values of x and y may be easily verified by substitution in the original equations.

Thus, $11\times8+3\times4=100$; or $88+12=100.$

And $4\times8-7\times4=\ \ 4$; or $32-28=\ \ 4.$

Ex. 3. Given $\left.\begin{aligned}\frac{x}{2}+\frac{y}{3}&=7\\ \frac{x}{3}+\frac{y}{2}&=8\end{aligned}\right\}$ to find the values of x and y.

Ans. $x=6$, $y=12$.

Ex. 4. Given $\left.\begin{array}{l}\frac{x+2}{3}+8y=31\\ \frac{y+5}{4}+10x=192\end{array}\right\}$ to find the values of x and y

Ex. 5. Given $\left.\begin{array}{l}2y-\frac{x+3}{4}=7+\frac{3x-2y}{5}\\ 4x-\frac{8-y}{3}=24\frac{1}{2}-\frac{2x+1}{2}\end{array}\right\}$ to find the values of x and y.

Ex. 6. Given $\left.\begin{array}{l}\frac{a}{x}+\frac{b}{y}=m\\ \frac{c}{x}+\frac{d}{y}=n\end{array}\right\}$ to find the values of x and y.

Ans. $x=\frac{bc-ad}{nb-md}$; $y=\frac{bc-ad}{mc-na}$.

(113.) When a problem involves a large number of quantities, it is common to designate a part of them by different letters, and for the remaining quantities to employ the same letters accented or numbered.

Thus, a, a', a'', a''', a'''' . . . . . . $a^{(m)}$

$a^{(1)}, a^{(2)}, a^{(3)}, a^{(4)}$ . . . . . . $a^{(m)}$

a_1, a_2, a_3, a_4 . . . . . . a_m

$a_{,}, a_{,,}, a_{,,,}, a_{,,,,}$ . . . . . . a_m

are used to denote different quantities, though they generally imply some connection between the quantities which they represent. a' is read *a prime;* a'', *a second;* a''', *a third,* &c. We must carefully distinguish between a_2 and a^2; between a_4 and a^4, &c. In the one case, the numerals are exponents, and denote powers of a; while in the other case, the numerals are only used for the sake of convenience to denote distinct quantities. Examples showing the convenience of this notation will be found in Sections XIX. and XX.

Ex. 7. Given $\left.\begin{array}{l}ax+by=c\\ a'x+b'y=c'\end{array}\right\}$ to find the values of x and y

Ans. $x=\frac{b'c-bc'}{ab'-a'b}$; $y=\frac{ac'-a'c}{ab'-a'b}$.

The symmetry of these expressions is well calculated to fix them in the memory.

Ex. 8. What fraction is that, to the numerator of which, if 4 be added, the value is one half; but if 7 be added to the denominator, its value is one fifth?

Let $\frac{x}{y}$ represent the fraction required.

Then, by the first condition,

$$\frac{x+4}{y}=\frac{1}{2}; \text{ whence } 2x+8=y.$$

By the second condition,

$$\frac{x}{y+7}=\frac{1}{5}; \text{ whence } 5x=y+7.$$

Subtracting the first equation from the second, we have

$$3x-8=7,$$

whence $3x=15,$

or $x=5.$

Therefore, $y=2x+8=10+8=18,$

and the fraction is $\frac{5}{18}$.

Proof. $\frac{5+4}{18}=\frac{1}{2}$

and $\frac{5}{18+7}=\frac{1}{5}.$

Ex. 9. A certain sum of money, put out at simple interest, amounts in 8 months to $1488, and in 15 months it amounts to $1530. What is the sum and rate per cent.?

Ex. 10. A sum of money put out at simple interest amounts in m months to a dollars, and in n months to b dollars.

Required the sum and rate per cent.?

Ans. The sum is $\frac{na-mb}{n-m}$; the rate is $1200\times\frac{b-a}{na-mb}$.

Ex. 11. There is a number consisting of two digits, the second of which is greater than the first; and if the number be divided by the sum of its digits, the quotient is 4; but if the digits be inverted, and that number be divided by a number greater by two than the difference of the digits, the quotient is 14.

Required the number.

Let x represent the left hand digit,

and y " right hand digit.

Then, since x stands in the place of tens, the number will be represented by $10x+y$.

Hence, by the first condition,

$$\frac{10x+y}{x+y}=4.$$

By the second condition,

$$\frac{10y+x}{y-x+2}=14.$$

Whence $x=4,\ y=8,$

and the required number is 48.

Ex. 12. A boy expends thirty pence in apples and pears, buying his apples at 4 and his pears at 5 for a penny, and afterward accommodates his friend with half his apples and one third of his pears for 13 pence. How many did he buy of each?

Ex. 13. A father leaves a sum of money to be divided among his children, as follows: the first is to receive \$300 and the sixth part of the remainder; the second \$600 and the sixth part of the remainder; and, generally, each succeeding one receives \$300 more than the one immediately preceding, together with the sixth part of what remains. At last it is found that all the children receive the same sum. What was the fortune left and the number of children?

Ans. The fortune was \$7500, the number of children 5.

Ex. 14. A sum of money is to be divided among several persons, as follows: the first receives a dollars together with the nth part of the remainder; the second $2a$ together with the nth part of the remainder; and each succeeding one a dollars more than the preceding, together with the nth part of the remainder; and it is found, at last, that all have received the same sum. What was the amount divided, and the number of persons?

Ans. The amount $=a(n-1)^2$, the number of persons $=n-1$.

EQUATIONS WHICH CONTAIN THREE OR MORE UNKNOWN QUANTITIES.

(114.) Let us now consider the case of *three* equations involving *three* unknown quantities.

Take the system of equations,

$$3x+2y+\ z=16, \qquad (1.)$$
$$2x+2y+2z=18, \qquad (2.)$$
$$2x+3y+\ z=17. \qquad (3.)$$

In order to eliminate z between equations (1.) and (2.), we will divide both members of the second equation by two; we thus obtain

$$x+y+z=9.$$

Subtracting this from the first equation, we find a new equation containing but two unknown quantities,

$$2x+y=7. \qquad (\alpha.)$$

In order to eliminate z between equations (1.) and (3.), subtract the former from the latter, which gives

$$-x+y=1. \qquad (\beta.)$$

From the two equations (α.) and (β.), one may be deduced containing only one unknown quantity. For, by subtracting the one from the other, we have

$$3x=6, \text{ or } x=2.$$

Substituting this value of x in equation (β.), we obtain

$$y=3.$$

Substituting these values of x and y in equation (1.), we obtain

$$3\times2+2\times3+z=16.$$

Hence $$z=4.$$

These values of x, y, and z may be verified by substitution in the original equations.

We have effected the elimination in this case by method third, *Art.* 111; but either of the other methods might have been employed. Hence, to solve three equations containing three unknown quantities, we have the following

RULE.

(115.) *From the three equations, deduce two containing only two unknown quantities; then from these two deduce one containing only one unknown quantity.*

Ex. 15. Given
$$\left.\begin{aligned} x+y+z&=29 \ (1.) \\ x+2y+3z&=62 \ (2.) \\ \tfrac{1}{2}x+\tfrac{1}{3}y+\tfrac{1}{4}z&=10 \ (3.) \end{aligned}\right\} \text{ to find } x, y, \text{ and } z.$$

Subtract equation (1.) from (2.), and we obtain

$$y+2z=33; \qquad (\alpha.)$$

clearing equation (3.) of fractions, we have

$$6x+4y+3z=120. \qquad (4.)$$

Multiplying equation (1.) by 6,

$$6x+6y+6z=174. \qquad (5.)$$

Subtracting (4.) from (5.), $2y+3z=54$. (β.)

We have thus obtained two equations, (α.) and (β.), containing two unknown quantities.

Multiplying (α.) by 2, we have $2y+4z=66$. (6.)

Subtracting (β.) from (6.), $z=12$.

Substituting this value of z in (β.), we obtain

$$2y+36=54.$$

Whence $y=9$.

Substituting these values of y and z in equation (1.),

$$x+9+12=29.$$

Whence $x=8$.

These values may be verified as in former examples.

Ex. 16. Given $\left.\begin{array}{l}2x+4y-3z=22\\4x-2y+5z=18\\6x+7y-\ z=63\end{array}\right\}$ to find x, y, and z.

Ans. $x=3$, $y=7$, $z=4$.

Ex. 17. Given $\left.\begin{array}{l}x+y=a\\x+z=b\\y+z=c\end{array}\right\}$ to find x, y, and z.

Ex. 18. Given $\left.\begin{array}{l}x+\frac{1}{2}y+\frac{1}{3}z=32\\\frac{1}{3}x+\frac{1}{4}y+\frac{1}{5}z=15\\\frac{1}{4}x+\frac{1}{5}y+\frac{1}{6}z=12\end{array}\right\}$ to find x, y, and z.

(116.) If we had *four* equations involving *four* unknown quantities, we might, by the methods already explained, eliminate one of the unknown quantities. We should thus obtain *three* equations between *three* unknown quantities, which might be solved according to *Art.* 114. So, also, if we had *five* equations involving *five* unknown quantities, we might, by the same process, reduce them to *four* equations involving *four* unknown quantities; then to *three*, and so on. By following the same method, we might resolve a system of any number of equations of the first degree. Hence, if we have m equations involving m unknown quantities, we proceed by the following

RULE.

1. *Combine successively any one of the equations with each of the others, so as to eliminate the same unknown quantity; we*

thus obtain m−1 *new equations containing* m−1 *unknown quantities.*

2. *Eliminate another unknown quantity by combining any one of these new equations with the others; there will result* m−2 *equations containing* m−2 *unknown quantities.*

3. *Continue this series of operations until there results a single equation containing but one unknown quantity, from which the value of this unknown quantity is easily deduced. Then by going back, step by step, to one of the original equations, the values of the other unknown quantities may be successively determined.*

Ex. 19. Given
$$\left.\begin{aligned} 7x-2z+3u&=17\\ 4y-2z+t&=11\\ 5y-3x-2u&=8\\ 4y-3u+2t&=9\\ 3z+8u&=33 \end{aligned}\right\}\text{ to find } x, y, z, u, \text{ and } t.$$

Ans. $x=2$, $y=4$, $z=3$, $u=3$, $t=1$.

Either of the unknown quantities may be selected as the one to be first exterminated. It is, however, generally best to begin with that which has the smallest coefficients; and if each of the unknown quantities is not contained in all the proposed equations, it is generally best to begin with that which is found in the least number of equations.

Ex. 20. A person owes a certain sum to two creditors. At one time he pays them \$530, giving to one four elevenths of the sum which is due, and to the other \$30 more than one sixth of his debt to him. At a second time he pays them \$420, giving to the first three sevenths of what remains due to him, and to the other one third of what remains due to him. What were the debts?

Ex. 21. If A and B together can perform a piece of work in 12 days, A and C together in 15 days, and B and C in 20 days, how many days will it take each person to perform the same work alone?

This Problem is readily solved by first finding in what time they could finish it if all worked together.

Ex. 22. If A and B together can perform a piece of work in *a* days, A and C together in *b* days, and B and C in *c* days, how many days will it take each person to perform the same work alone?

Ans. A requires $\frac{2abc}{ac+bc-ab}$ days,

B " $\frac{2abc}{ab+bc-ac}$ days,

C " $\frac{2abc}{ab+ac-bc}$ days.

(117.) Hitherto we have supposed the number of equations equal to the number of symbols employed to denote the unknown quantities. This must be the case with every problem, in order that it may be *determinate;* that is, that it may not admit of an indefinite number of solutions.

Suppose, for example, that a problem involving two unknown quantities (x and y) leads to the single equation

$$x-y=3.$$

Now if we make $y=1$, then $x=4$;
$y=2$, then $x=5$;
$y=3$, then $x=6$;
$y=4$, then $x=7$,
&c., &c.;

and each of these systems of values, 1 and 4, 2 and 5, 3 and 6 &c., substituted for x and y in the original equation, will satisfy it equally well. Hence the problem is *indeterminate;* that is, admits of an indefinite number of solutions.

(118.) If we had *two* equations involving *three* unknown quantities, we could, in the first place, eliminate one of the unknown quantities by means of the proposed equations, and thus obtain *one* equation containing *two* unknown quantities, which would be satisfied by an infinite number of systems of values. Therefore, in order that a problem may be *determinate,* its enunciation must contain as many different conditions as there are unknown quantities, and each of these conditions must be expressed by an *independent* equation.

Equations are said to be *independent* when they express conditions *essentially different;* and *dependent* when they express the *same* conditions under different forms.

Thus, $\left.\begin{aligned} x+y&=7 \\ 2x+y&=10 \end{aligned}\right\}$ are *independent* equations.

But $\left.\begin{array}{r} x+\ y=\ 7 \\ 2x+2y=14 \end{array}\right\}$ are *not* independent,

because the one may be deduced from the other.

(119.) If, on the contrary, the number of *independent* equations exceeds the number of unknown quantities, these equations will be *contradictory*.

For example, let it be required to find two numbers such that their sum shall be 7, their difference 1, and their product 100.

From these conditions we derive the following equations.

$$x+y=\ \ 7,$$
$$x-y=\ \ 1,$$
$$xy=100.$$

From the first two equations we easily find

$$x=4, \text{ and } y=3.$$

Hence the third condition, which requires that their product shall be equal to 100, *can not be fulfilled.*

SECTION IX.

DISCUSSION OF EQUATIONS OF THE FIRST DEGREE. INEQUALITIES.

(120.) To *discuss* a problem or an equation is to determine the values which the unknown quantities assume for particular hypotheses made upon the values of the given quantities, and to interpret the peculiar results obtained. The term, therefore, is not strictly applicable, except to problems which are stated in the most general form, like some of those in *Arts.* 106 and 107. If the sum of two numbers is represented by a and their difference by b, the greater number will be expressed by $\frac{a}{2}+\frac{b}{2}$, and the less by $\frac{a}{2}-\frac{b}{2}$. Here a and b may have any values whatever, and still these formulæ will always hold true. It frequently happens that, by attributing different values to the letters which represent known quantities, the values of the unknown quantities assume peculiar forms which deserve consideration.

(121.) We may obtain five species of values for the unknown quantity in a problem of the first degree.

I. Positive values.

II. Negative values.

III. Values of the form of zero, or $\frac{0}{A}$.

IV. Values of the form of $\frac{A}{0}$.

V. Values of the form of $\frac{0}{0}$.

We will consider these five cases in succession.

I. *Positive values* are generally answers to problems in the sense in which they are proposed. Nevertheless, all positive values will not always satisfy the enunciation of a problem. If, for example, a problem requires an answer in whole numbers, and we obtain a fractional value, the problem is impossible. Thus, in Problem 17, page 71, it is implied that the value of x must be a whole number, although this condition is not expressed in the equations. It would be easy to change the data of the problem so as to obtain a fractional value of x, which would indicate an impossibility in the problem proposed. Problem 43, page 76, is of the same kind; also *Ex.* 11, page 85.

If the value obtained for the unknown quantity, even when positive, does not satisfy all the conditions of the problem, the problem is impossible in the form proposed.

(122.) II. *Negative values.*

Let it be proposed to find a number, which, added to the number b, gives for a sum the number a.

Let $x =$ the required number.

Then, by the terms of the problem,

$$x+b=a, \text{ whence } x=a-b.$$

This formula will give the value of x for every case of the proposed problem.

For example, let $a=7$, and $b=4$.

Then $x=7-4=3$.

Again, let $a=5$, and $b=8$.

Then $x=5-8=-3$.

We thus obtain for x a *negative* value. How is it to be interpreted?

By referring to the problem, we see that it is proposed to find a number which, added to 8, shall make it equal to 5. Considered arithmetically, the problem is plainly impossible. Nevertheless, if in the equation $8+x=5$, we substitute for $+x$ its value -3, it becomes

$$8-3=5,$$

an identical equation; that is, 8 diminished by 3 is equal to 5.

The negative solution $x=-3$, shows, therefore, the impossibility of satisfying the enunciation of the problem as above stated; but, taking this value of x with a contrary sign, we see that it satisfies the enunciation when modified as follows:

To find a number which, *subtracted* from 8, gives a *difference* of 5; an enunciation which differs from the former only in this, that we put *subtract* for *add*, and *difference* for *sum*.

If we wish to solve this new question directly, we shall have

$$8-x=5.$$

Whence $x=8-5$, or $x=3$.

(123.) For another example, take Problem 50, page 77 The age of the father being represented by a, and that of the son by b; then $\frac{a-nb}{n-1}$ will represent the number of years before the age of the father will be n times that of the son.

Thus, suppose $a=54$, $b=9$, and $n=4$.

Then $x=\frac{54-36}{3}=\frac{18}{3}=6.$

That is, the father having lived 54 years and the son 9, in 6 years more the father will be 60 years old and the son 15. But 60 is 4 times 15; hence this value, $x=6$, satisfies the enunciation of the problem.

Again, suppose $a=45$, $b=15$, and $n=4$.

Then $x=\frac{45-60}{3}=\frac{-15}{3}=-5.$

Here again we obtain a *negative solution*. How are we to interpret it?

By referring to the problem, we see that the age of the son is already *more* than one fourth that of the father, so that the time required is already *past* by five years. The value of x just obtained, taken with a contrary sign, satisfies the following enunciation:

A father is 45 years old, his son 15; *how many years since* the age of the father was four times that of his son?

The equation corresponding to this new enunciation is

$$15-x=\frac{45-x}{4}.$$

Whence $60-4x=45-x$; and $x=5$.

(124.) Reasoning from analogy, we deduce the following general principles:

1. *Every negative value found for the unknown quantity in a*

problem of the first degree, indicates an absurdity in the conditions of the problem, or at least in its algebraic statement.

2. *This value, taken with a contrary sign, may be regarded as the answer to a problem, whose enunciation only differs from that of the proposed problem in this, that certain quantities which were* ADDED *should have been* SUBTRACTED, *and reciprocally.*

(125.) In what case would the value of the unknown quantity in Prob. 20, page 72, be negative?

Ans. When $n>m$.

Thus, let $m=20$, $n=25$, and $a=60$ miles.

Then $$x=\frac{60}{20-25}=\frac{60}{-5}=-12.$$

To interpret this result, observe that it is impossible that the second train, which moves the slowest, should overtake the first. At the time of starting, the distance between them was 60 miles, and every subsequent hour the distance increases. If, however, we suppose the two trains to *have been* moving uniformly along an endless road, it is obvious that *at some former time they must have been together.*

This negative solution then shows an absurdity in the conditions of the problem. The problem should have been stated thus:

Two trains of cars, 60 miles apart, are moving in the same direction, the forward one 25 miles per hour, the other 20. *How long since they were together?*

To solve this problem, let x = the required number of hours
Then $25x$ = the distance traveled by the first train,
$20x$ = " " second train.

And since they are now 60 miles apart,

$$25x=20x+60.$$

Hence $$5x=60,$$
and $$x=+12.$$

We thus obtain a positive value of x.

In order to include both of these cases in the same enunciation, the question should have been asked, *Required the time of their being together*, leaving it uncertain whether the time was *past* or *future.*

In what case would the value of one of the unknown quan

tities in Problem 34, page 74, be negative? Why should it be negative? and how could the enunciation be corrected for this case?

In what case would the value of one of the unknown quantities in Problem 4, page 67, be negative?

(126.) III. *Values of the form of zero, or* $\frac{0}{A}$.

In what case would the value of the unknown quantity in Problem 20, page 72, become zero, and what would this value signify?

Ans. This value becomes zero when $a=0$, which signifies that the two trains are together at the outset.

In what case would the value of the unknown quantity in Problem 50, page 77, become zero, and what would this value signify?

Ans. When $a=nb$, which signifies that the age of the father is *now* n times that of the son.

In what case would the values of the unknown quantities in Problem 38, page 75, become zero, and what would this signify?

When a problem gives zero for the value of the unknown quantity, this value is sometimes applicable to the problem, and sometimes it indicates an impossibility in the proposed question.

(127.) IV. *Values of the form of* $\frac{A}{0}$.

In what case does the value of the unknown quantity in Problem 20, page 72, reduce to $\frac{A}{0}$? and how shall we interpret this result?

Ans. When $m=n$.

On referring to the enunciation of the problem, we see that it is absolutely impossible to satisfy it; that is, there can be no point of meeting, for the two trains being separated by the distance a, and moving equally fast, will always continue at the same distance from each other. The result $\frac{a}{0}$ may then be regarded as indicating an *impossibility*.

The symbol $\frac{a}{0}$ is sometimes employed to represent *infinity;* and for the following reason:

When the difference $m-n$, without being absolutely nothing, is *very small*, the quotient $\frac{a}{m-n}$ is *very large.*

For example, let $m-n=0.01$.

Then $$x=\frac{a}{m-n}=\frac{a}{.01}=100a.$$

Let $$m-n=0.0001,$$

$$\frac{a}{m-n}=\frac{a}{.0001}=10000a.$$

Hence, if the difference in the rates of motion is not zero, the two trains must meet, and the time will become greater and greater as this difference is diminished. *If, then, we suppose this difference less than any assignable quantity, the time represented by $\frac{a}{m-n}$ will be greater than any assignable quantity, or infinite.*

Hence we infer, that every expression of the form $\frac{A}{0}$, found for the unknown quantity, indicates the impossibility of satisfying the problem, at least in *finite* numbers.

In what case would the value of the unknown quantity in Problem 10, page 70, reduce to the form $\frac{A}{0}$? and how shall we interpret this result?

(128.) V. *Values of the form of* $\frac{0}{0}$.

In what case does the value of the unknown quantity in Problem 20, page 72, reduce to $\frac{0}{0}$? and how shall we interpret this result?

Ans. When $a=0$, and $m=n$.

To interpret this result, let us recur to the enunciation, and observe that, since a is zero, both trains start from the same point; and since they both travel at the same rate, *they will always remain together*, and therefore the required point of meeting will be any where in the road traveled over. Th problem, then, is entirely *indeterminate*, or admits of an infinite number of solutions, and the expression $\frac{0}{0}$ may represent any finite quantity.

We infer, therefore, that an expression of the form $\frac{0}{0}$ found for the unknown quantity, generally indicates that it may have any value whatever. In some cases, however, this value is subject to limitations.

In what case would the values of the unknown quantities in Problem 44, page 76, reduce to $\frac{0}{0}$? and how would they satisfy the conditions of the problem?

Ans. When $a=b=c$,

which indicates that the coins are all of the same value. B might therefore be paid in either kind of coin; but there is a limitation, viz., that the value of the coins must be one dollar.

In what case do the values of the unknown quantities in Problem 38, page 75, reduce to $\frac{0}{0}$? and how shall we interpret this result?

OF ZERO AND INFINITY.

(129.) From *Art.* 127, it is seen that in Algebra we sometimes have occasion to consider infinite quantities. It is necessary, therefore, to establish some general principles respecting them.

An infinite quantity is one which exceeds any assignable limit. It is often expressed by the character ∞. Thus, a line produced beyond any assignable limit is said to be of infinite length. A surface indefinitely extended, and also a solid of indefinite extent in any one of its three dimensions, are examples of infinity.

An infinite quantity does not mean an infinite *number of terms.* Thus, the fraction $\frac{1}{3}$ reduced to a decimal, is .333333, &c., without end, but the value of this series is less than unity.

Infinite quantities are *not all equal* among themselves.

Thus the series $1+1+1+1+1+$, &c.,
$2+2+2+2+2+$, &c.,
$3+3+3+3+3+$, &c.,

continued to an infinite number of terms, will each be infinite, although the second series will be double, and the third treble the first.

So, also, a line may be infinitely extended both ways; or it may be infinitely extended in one direction, and limited in the other. In either case, the line is said to be infinite.

A quantity less than any assignable quantity is called an infinitesimal, and is sometimes represented by 0.

Thus, take the series of fractions $\frac{1}{10}$, $\frac{1}{100}$, $\frac{1}{1000}$, $\frac{1}{10000}$, &c. By increasing the denominator, we diminish the value of the fraction; and if the denominator be made infinitely great, the quotient will be infinitely small.

(130.) We have seen, in *Art.* 127, that $\frac{a}{0}=\infty$, where a may represent any finite quantity. That is,

If a finite quantity be divided by zero, the quotient is infinite

From the same equation we deduce $\frac{a}{\infty}=0$. That is,

If a finite quantity be divided by infinity, the quotient is zero

From the same equation we deduce $a=0\times\infty$. That is,

If zero be multiplied by infinity, the product is a finite quantity.

If a finite quantity be multiplied by a proper fraction, it will be *diminished*, and the smaller the multiplier, the less the product. Hence, if the multiplier be infinitely small, the product will be infinitely small, or $a\times 0=0$. That is,

If a finite quantity be multiplied by zero, the product will be zero.

From this equation we deduce $a=\frac{0}{0}$; that is,

If zero be divided by zero, the quotient may be any finite quantity.

The greater the multiplier, the greater will be the product. Hence, *if a finite quantity be multiplied by infinity, the product will be infinite;* that is,

$$a\times\infty=\infty.$$

From this equation we deduce $a=\frac{\infty}{\infty}$; that is,

If infinity be divided by infinity, the quotient may be any finite quantity.

An infinite quantity can not be increased by the addition of a finite quantity, or diminished by its subtraction; that is, $\infty \pm a = \infty$.

So, also, a finite quantity is not altered by the addition or subtraction of zero; that is, $a \pm 0 = a$.

OF INEQUALITIES.

(131.) In discussing algebraical problems, as shown in *Arts.* 120–128, it is frequently necessary to employ *inequalities*, or expressions of two quantities which are *not* equal to each other. Generally, the principles already established for the transformation of equations are applicable to inequalities also. There are, however, some important exceptions to be noted, arising chiefly from the use of *negative expressions* as *quantities.*

Two inequalities are said to subsist in *the same sense* when the greater quantity stands at the left in both, or at the right in both; and in *a contrary sense* when the greater quantity stands at the right in one, and at the left in the other.

Thus, $9>7$ and $7>6$.

As also $5<8$ and $3<4$,

are inequalities which subsist in the same sense; but the inequalities

$$10>6 \text{ and } 3<7,$$

subsist in a contrary sense.

(132.) I. *If we add the same quantity to both members of an inequality, or subtract the same quantity from both members, the resulting inequality will always subsist in the same sense.*

Thus, $8>3$.

Adding 5 to each member,

$$8+5>3+5;$$

and subtracting 5 from each member,

$$8-5>3-5.$$

Again, take the inequality $-3<-2$.

Adding 6 to each member, we have

$$-3+6<-2+6, \text{ or } 3<4;$$

and subtracting 6 from each member,

$$-3-6<-2-6, \text{ or } -9<-8.$$

The student must here bear in mind what was stated in *Art.* 47, of two negative quantities, that is the *least* whose numerical value is the *greatest.*

This principle enables us to *transpose* any term from one member of an inequality to the other by changing its sign

Thus, $a^2+b^2>3b^2-2a^2.$

Adding $2a^2$ to each member of the inequality, it becomes

$$a^2+b^2+2a^2>3b^2.$$

Subtracting b^2 from each member,

$$a^2+2a^2>3b^2-b^2,$$

or

$$3a^2>2b^2.$$

(133.) II. *If we add together the corresponding members of two or more inequalities which subsist in the same sense, the resulting inequality will always subsist in the same sense.*

Thus,

$$\begin{array}{r} 5>4 \\ 4>2 \\ 7>3 \\ \hline \end{array}$$

Adding, we obtain $16>9.$

III. *But if we subtract the corresponding members of two or more inequalities which subsist in the same sense, the resulting inequality will* NOT ALWAYS *subsist in the same sense.*

Take the inequalities

$$\begin{array}{r} 4<7 \\ 2<3 \\ \hline \end{array}$$

Subtracting, we have $4-2<7-3$, or $2<4$, where the resulting inequality subsists in the *same* sense.

But take $9<10$

and $6<8$.

Subtracting, the result is $9-6>$ (not $<$) $10-8$, or $3>2$, where the resulting inequality subsists in the *contrary* sense.

We should therefore avoid as much as possible the use of this transformation, or when we employ it, determine in what sense the resulting inequality subsists.

(134.) IV. *If we multiply or divide the two members of an inequality by a positive number, the resulting inequality will subsist in the same sense.*

Thus, if $a < b$.

Then $ma < mb$.

And $\frac{a}{m} < \frac{b}{m}$.

Also, if $-a > -b$.

Then $-na > -nb$.

And $-\frac{a}{n} > -\frac{b}{n}$.

This principle will enable us to clear an inequality of fractions. Thus, suppose we have

$$\frac{a^2-b^2}{2d} > \frac{c^2-d^2}{3a}.$$

Multiplying both members by $6ad$, it becomes

$$3a(a^2-b^2) > 2d(c^2-d^2).$$

V. *If we multiply or divide the two members of an inequality by a negative number, the resulting inequality will subsist in a contrary sense.*

Take, for example, $8 > 7$.

Multiplying both members by -3, we have the opposite inequality,

$$-24 < -21.$$

So, also, $15 > 12$.

Dividing each member by -3, we have

$$-5 < -4.$$

Therefore, if we multiply or divide the two members of an inequality by an algebraic quantity, it is necessary to ascertain whether the multiplier or divisor is negative, for in this case the inequality subsists in a contrary sense.

VI. *If we change the signs of both members of an inequality, we must reverse the sense of the inequality,* for this transformation is evidently the same as multiplying both members by -1.

(135.) VII. *If both members of an inequality are positive numbers, we can raise them to any power without changing the sense of the inequality.*

Thus, $5 > 3$,

so also, $5^2 > 3^2$, or $25 > 9$.

And if $a > b$,
then will $a^n > b^n$.

VIII. *If both members of an inequality are not positive numbers, and they be raised to any power, the resulting inequality will not always subsist in the same sense.*

Thus, $-2 < +3$,
gives $(-2)^2 < 3^2$, or $4 < 9$,
where the resulting inequality subsists in the *same* sense.

But $-3 > -5$,
gives $(-3)^2 < (-5)^2$, or $9 < 25$,
where the resulting inequality subsists in a *contrary* sense.

IX. *In extracting the root of both members of an inequality, it is sometimes necessary to reverse the sense of the inequality.*

Thus, from $9 < 25$,
by extracting the square root, we obtain
either $3 < 5$,
or $-3 > -5$.

EXAMPLES.

1. Given $7x-3<25$, to find the limit of x.

Ans. $x<4$.

2. Given $2x+\frac{x}{3}-8<6$, to find the limit of x.

Ans. $x<6$.

3. Given $\frac{x}{2}+\frac{x}{3}+\frac{x}{4}+\frac{x}{6}+\frac{x}{12}-7>9$, to find the limit of x.

4. Given $\left.\begin{array}{l}\frac{bx}{2}+cx-ac<\frac{ab}{2}\\ ex+\frac{dx-bd}{3}>be\end{array}\right\}$, to find the limits of x.

5. A man being asked how many dollars he gave for his watch, replied, If you multiply the price by 4, and to the product add 60, the sum will exceed 256; but if you multiply the price by 3, and from the product subtract 40, the re-

mainder will be less than 113. Required the price of the watch.

6. What number is that whose half and third part added together are less than 105, but its half diminished by its fifth part is greater than 33?

7. The double of a number diminished by 6 is greater than 24, and triple the number diminished by 6 is less than double the number increased by 10. Required the number.

SECTION X.

INVOLUTION AND POWERS.

(136.) According to *Art.* 20, *the products formed by the successive multiplication of the same number by itself are called the powers of that number.*

Thus, the first power of 3 is 3.

The second power of 3 is 9, or 3×3.

The fourth power of 3 is 81, or $3\times3\times3\times3$,

&c., &c., &c.

According to *Art.* 21, *the exponent is a number or letter written a little above a quantity to the right, and denotes the number of times that quantity enters as a factor into a product.*

Thus, the first power of a is a^1, where the exponent is 1, which, however, is commonly omitted.

The second power of a is $a\times a$, or a^2, where the exponent 2 denotes that a is taken twice as a factor to produce the power aa.

The third power of a is $a\times a\times a$, or a^3, where the exponent 3 denotes that a is taken three times as a factor to produce the power aaa.

The fourth power of a is $a\times a\times a\times a$, or a^4.

Also, the nth power of a is $a\times a\times a\times a$ . . . repeated as a factor n times, and is written a^n.

Exponents may be applied to polynomials as well as to monomials.

Thus $(a+b+c)^3$ is the same as

$$(a+b+c)\times(a+b+c)\times(a+b+c),$$

or the third power of the entire expression $a+b+c$.

(137.) According to the rule for the multiplication of monomials, *Arts.* 49 and 50.

$$(3ab^2)^2=3ab^2\times 3ab^2=9a^2b^4.$$

So, also, $(4a^2bc^3)^2=4a^2bc^3\times 4a^2bc^3=16a^4b^2c^6.$

Hence it appears that, *in order to square a monomial, we must square its coefficient, and multiply the exponent of each of the letters by* 2.

EXAMPLES.

1. Required the square of $7axy$.

Ans. $49a^2x^2y^2$.

2. Required the square of $11a^3bcd^2$.
3. Required the square of $12a^2xy$.
4. Required the square of $15ab^2cx^4$.
5. Required the square of $18x^2yz^3$.

According to *Art.* 53, $+$ multiplied by $+$, and $-$ multiplied by $-$, give $+$. Now the square of any quantity being the product of that quantity by itself, it necessarily follows that *whatever may be the sign of a monomial, its square must be affected with the sign* $+$.

Thus the square of $+3ax$ or of $-3ax$ is $+9a^2x^2$.

(138.) The method of involving a quantity to *any* power, is easily derived from the preceding principles.

Let it be required to form the fifth power of $2a^3b^2$.

According to the rules for multiplication,

$$(2a^3b^2)^5=2a^3b^2\times 2a^3b^2\times 2a^3b^2\times 2a^3b^2\times 2a^3b^2$$
$$=32a^{15}b^{10}.$$

Where we perceive

1. That the coefficient has been raised to the fifth power.
2. That the exponent of each of the letters has been multiplied by 5.

In like manner,

$$(3a^2b^3c)^3=3a^2b^3c\times 3a^2b^3c\times 3a^2b^3c$$
$$=3^3a^{2+2+2}b^{3+3+3}c^{1+1+1}$$
$$=27a^6b^9c^3.$$

Hence, to raise a monomial to any power, we have the following

RULE.

Raise the numerical coefficient to the given power, and multiply the exponent of each of the letters by the exponent of the power required.

EXAMPLES.

1. Required the fourth power of $4ab^2c^3$.

Ans. $256a^4b^8c^{12}$.

2. Required the fifth power of $3ax^3y^5$.
3. Required the third power of $6xy^2z^4$.
4. Required the sixth power of $2ad^2y^3v$.
5. Required the seventh power of $2a^2bc^4$.
6. Required the sixth power of $5w^3xy^2z^4$.

(139.) Let us now consider the sign with which the power should be affected.

We have seen, *Art.* 137, that whatever may be the sign of a monomial, its square is always positive. It is obvious, from the same considerations, that the product of an *even* number of negative factors is *positive*, but the product of an *odd* number of negative factors is *negative*.

Thus,

$$-a\times-a=+a^2$$
$$-a\times-a\times-a=-a^3$$
$$-a\times-a\times-a\times-a=+a^4$$
$$-a\times-a\times-a\times-a\times-a=-a^5$$

&c., &c., &c.

The product of several factors which are all *positive*, is invariably positive. Hence,

Every EVEN *power is positive, but an* ODD *power has the same sign as its root.*

EXAMPLES.

1. Required the square of $-2x^5$.

Ans. $+4x^{10}$.

2. Required the square of $-3x^n$.
3. Required the cube of $-3a^3$.
4. Required the fourth power of $-3a^3b^2h$.
5. Required the fifth power of $-2a^3\times3x^2y$.

(140.) *A fraction is involved by involving both the numerator and denominator.*

1. Thus, the square of $\frac{a}{b}$ is $\frac{a}{b}\times\frac{a}{b}$; which, by *Art.* 89, is equal to $\frac{a^2}{b^2}$, which, by *Art.* 68, may be written a^2b^{-2}.

2. Required the cube of $\frac{2ab^2}{3c}$.

Ans. $\frac{8a^3b^6}{27c^3}$, or $\frac{8}{27}a^3b^6c^{-3}$.

3. Required the nth power of $\frac{a^2b}{ay^m}$.

(141.) Hence, expressions with *negative* exponents are involved by the same rule as those with positive exponents.

Thus, let it be required to find the square of a^{-3}.

This expression may be written $\frac{1}{a^3}$, which, raised to the second power, becomes $\frac{1}{a^6}$ or a^{-6}, the same result as would be obtained by multiplying the exponent -3 by 2.

Ex. 1. Required the square of $3a^2b^{-4}$.

Ex. 2. Required the square of $7a^{-2}b^3c^{-4}dx^{-1}$.

Ex. 3. Required the cube of $-6ab^{-5}dy^{-2}$.

Ex. 4. Required the fourth power of $3a^{-n}b$.

Ex. 5. Required the fifth power of $-2ab^{-3}c^2$.

(142.) *A polynomial is involved by multiplying it into itself as many times less one as is denoted by the exponent of the power.*

Ex. 1. Required the fourth power of $a+b$.

$$\begin{array}{l} a+b \\ a+b \\ \hline a^2+ab \\ \quad +ab+b^2 \\ \hline \end{array}$$

$(a+b)^2=a^2+2ab+b^2$, the second power of $a+b$.

$$\begin{array}{l} a+b \\ \hline a^3+2a^2b+ab^2 \\ \quad +\ a^2b+2ab^2+b^3 \\ \hline \end{array}$$

$(a+b)^3=a^3+3a^2b+3ab^2+b^3$, the third power.

$$\begin{array}{l} a+b \\ \hline a^4+3a^3b+3a^2b^2+\ ab^3 \\ \quad +\ a^3b+3a^2b^2+3ab^3+b^4 \\ \hline \end{array}$$

$(a+b)^4=a^4+4a^3b+6a^2b^2+4ab^3+b^4$, the fourth power.

Ex. 2. Required the fourth power of $a-b$.

Ans. $a^4-4a^3b+6a^2b^2-4ab^3+b^4$.

Ex. 3. Required the cube of $2a-1$.

Ex. 4. Required the fourth power of $3a-h$.

Ex. 5. Required the square of $a+b+c$.

Hence it appears that *the square of a trinomial is composed of the sum of the squares of all the terms, together with twice the sum of the products of all the terms multiplied together two and two.*

Ex. 6. Required the cube of $2ab+cd$.

Ex. 7. Required the fourth power of a^3+b^3.

Ex. 8. Required the cube of $a+\frac{1}{2}$.

Ex. 9. Required the cube of $x+\frac{y}{2}$.

Ex. 10. Required the square of $a+b+c+d+e$.

From this example we infer that the *square of any polynomial is composed of the sum of the squares of all the terms, together with twice the sum of the products of all the terms multiplied together two and two,* and this proposition may be rigorously demonstrated.

It is obvious that this rule for a polynomial includes the preceding rule for a trinomial, and that in *Art.* 60 for a binomial.

Ex. 11. Required the fourth power of $2x-3y$.

Ans. $16x^4-96x^3y+216x^2y^2-216xy^3+81y^4$.

Ex. 12. Required the square of $a+m-n$.

Ans. $a^2+2am-2an+m^2-2mn+n^2$.

Ex. 13. Required the cube of $a+b-x$.

Ans. $a^3+b^3-x^3+3ab^2+3ax^2+3a^2b-3a^2x+3bx^2-3b^2x-6abx$.

Ex. 14. Required the cube of $\frac{2a+3b}{m-n}$.

Ans. $\frac{8a^3+36a^2b+54ab^2+27b^3}{m^3-3m^2n+3mn^2-n^3}$.

Ex. 15. Required the square of $\frac{ax-by}{ay-bx}$.

Ans. $\frac{a^2x^2-2abxy+b^2y^2}{a^2y^2-2abxy+b^2x^2}$.

Ex. 16. Required the cube of $\frac{a^2-b}{a-b^2}$.

Ans. $\frac{a^6-3a^4b+3a^2b^2-b^3}{a^3-3a^2b^2+3ab^4-b^6}$.

SECTION XI.

EVOLUTION AND RADICAL QUANTITIES.

(143.) *The square root of a quantity is a factor which, multiplied by itself once, will produce that quantity.*

Thus, the square root of a^2 is a, because a when multiplied by itself produces a^2.

The square root of 144 is 12 for the same reason.

According to *Art.* 22, the square root is indicated by the sign $\sqrt{\ }$.

Thus, $$\sqrt{a^2}=a,$$
and $$\sqrt{144a^2}=12a.$$

(144.) According to *Art.* 137, in order to square a monomial, we must square its coefficient, and multiply the exponent of each of its letters by 2. Therefore, in order to derive the square root of a monomial from its square, we must

I. *Extract the square root of its coefficient.*

II. *Divide each of the exponents by* 2.

Thus we shall have

$$\sqrt{64a^6b^4}=8a^3b^2.$$

This is manifestly the true result, for

$$(8a^3b^2)^2=8a^3b^2\times8a^3b^2=64a^6b^4.$$

So, also,

$$\sqrt{625a^2b^8c^6}=25ab^4c^3.$$

For, $$(25ab^4c^3)^2=25ab^4c^3\times25ab^4c^3,$$
$$=625a^2b^8c^6.$$

1. Required the square root of $196a^2b^4c^6d^8$.

2. Required the square root of $225a^{2m}b^{10}x^4$.

(145.) According to *Art.* 140, a fraction is involved by involving both the numerator and denominator; hence it is obvious that *the square root of a fraction is equal to the root of the numerator divided by the root of the denominator.*

Thus the square root of $\frac{a^2}{b^2}$ is $\frac{a}{b}$.

1. Find the square root of $\frac{4a^2}{9x^4y^4}$.

2. Find the square root of $\frac{9a^2b^4}{16c^6d^{12}}$.

(146.) It appears, from *Art.* 144, that *a monomial can not be a perfect square unless its coefficient be a square number, and the exponents of its letters all even numbers.*

Thus, $7ab^2$ is not a perfect square, for 7 is not a *square* number, and the exponent of a is not an *even* number. Its square root may be *indicated* by the usual sign, thus, $\sqrt{7ab^2}$. Expressions of this nature are called *surds,* or *radicals of the second degree.*

(147.) We have seen, *Art.* 137, that whatever may be the sign of a monomial, its square must be affected with the sign +. Hence we conclude that

If a monomial be positive, its square root may be either positive or negative.

Thus, $\sqrt{9a^4}=+3a^2$, or $-3a^2$,

for either of these quantities, when multiplied by itself, produces $9a^4$. We therefore always affect the square root of a quantity with the double sign $\pm$, which is read *plus* or *minus.*

Thus, $\sqrt{4a^6}=\pm 2a^3$

$\sqrt{25a^2b^4}=\pm 5ab^2$.

(148.) If a monomial be affected with a *negative* sign, the extraction of its square root is *impossible,* since we have just seen that the square of every quantity, whether positive or negative, is necessarily positive.

Thus, $\sqrt{-4}$, $\sqrt{-9}$, $\sqrt{-5a}$,

are algebraic symbols representing operations which it is impossible to execute. Quantities of this nature are called *imaginary* or *impossible* quantities, and are symbols of absurdity which we frequently meet with in resolving quadratic equations.

Such quantities may be represented by the form

$$\sqrt{-a}, \text{ which equals}$$
$$\sqrt{a\times-1}=\sqrt{a}\sqrt{-1}.$$

So that $\sqrt{a}\sqrt{-1}$ is a general form for all imaginary quantities of the second degree. Thus,

$$\sqrt{-4}=\sqrt{4\times-1}=2\sqrt{-1},$$
$$\sqrt{-9}=\sqrt{9\times-1}=3\sqrt{-1},$$
$$\sqrt{-5a}=\sqrt{5a\times-1}=\sqrt{5a}\sqrt{-1}.$$

That is, *the square root of a negative quantity may always be represented by the square root of a positive quantity multiplied by the square root of* -1.

(149.) According to *Art.* 138, in order to raise a monomial to *any* power, we raise the numerical coefficient to the given power, and multiply the exponent of each of the letters by the exponent of the power required. Hence, reciprocally, to extract any root of a monomial, we obtain the following

RULE.

I. *Extract the root of the numerical coefficient.*

II. *Divide the exponent of each letter by the index of the required root.*

Thus,
$$\sqrt[3]{64a^6b^3}=4a^2b.$$
$$\sqrt[4]{16b^{12}c^{16}}=2b^3c^4.$$

From *Art.* 145, it is obvious that *to extract* ANY *root of a fraction, we must divide the root of the numerator by the root of the denominator.*

Thus the cube root of $\frac{27a^6b^3}{8x^3y^9}$ is $\frac{3a^2b}{2xy^3}$;

which may be written $\frac{3}{2}a^2bx^{-1}y^{-3}$.

(150.) Let us now consider the *sign* with which the root should be affected. We have seen, *Art.* 139, that every *even* power is positive, but an *odd* power has the same sign as its root.

Thus $-a$, when raised to different powers in succession will give

$$-a, +a^2, -a^3, +a^4, -a^5, +a^6, -a^7, \&c.;$$

and $+a$, in like manner, will give

$$+a, +a^2, +a^3, +a^4, +a^5, +a^6, +a^7, \&c.$$

Since every even number may be expressed by $2n$, every even power may be considered as the square of the nth power, or $a^{2n}=(a^n)^2$, and must, therefore, be positive; and, in like manner, since an odd number may be expressed by $2n+1$, every power of an uneven degree may be considered as the product of the $2n$th power by the original quantity, and must, therefore, have the same sign with the monomial.

Hence it appears,

I. *An odd root of any quantity must have the same sign as the quantity itself.*

Thus,
$$\sqrt[3]{+8a^3}=+2a.$$
$$\sqrt[3]{-8a^3}=-2a.$$
$$\sqrt[5]{-32a^{10}b^5}=-2a^2b.$$
$$\sqrt[5]{+32a^{10}b^5}=+2a^2b.$$

II. *An even root of a positive quantity is ambiguous.*

Thus,
$$\sqrt[4]{81a^4b^{12}}=\pm 3ab^3.$$
$$\sqrt[6]{64a^{18}}=\pm 2a^3.$$

III. *An even root of a negative quantity is impossible.*

For no quantity can be found which, when raised to an even power, can give a negative result.

Thus, $\sqrt[4]{-a}$, $\sqrt[8]{-b}$, are symbols of operations which can not be performed, and they are therefore called *impossible* or *imaginary* quantities, as $\sqrt{-a}$, in *Art.* 148.

EXAMPLES.

1. Find the fourth root of $81a^8$. *Ans.* $\pm 3a^2$
2. Find the fifth root of $-243a^{10}b^5c^{-15}$.
3. Find the cube root of $-125a^3x^{-6}y^9$.
4. Find the square root of $\dfrac{16a^4}{9x^2y^2}$.
5. Find the fifth root of $\dfrac{32a^{10}b^{-5}}{243}$.

(151.) According to the rule of *Art.* 149, we perceive that, in order that a monomial may be a perfect power of any degree, its coefficient must be a perfect power of that degree, and the exponent of each letter must be divisible by the index of the root.

When the quantity whose root is required is *not* a perfect power of the given degree, we can only *indicate* the operation to be performed. Thus, if it be required to extract the cube root of $4a^2b^6$, the operation may be indicated by writing the expression thus,

$$\sqrt[3]{4a^2b^6}.$$

Expressions of this nature are called *surds,* or *irrational quantities,* or *radicals of the second, third,* or *n*th *degree,* according to the index of the root required.

(152.) The method of extracting the roots of polynomials will be considered in Section XVII. There is, however, one class so simple and of so frequent occurrence that it may properly be introduced here. In *Arts.* 60 and 61 we have seen that

the square of $a+b$ is $a^2+2ab+b^2$,
and the square of $a-b$ is $a^2-2ab+b^2$.

Therefore, the square root of $a^2\pm2ab+b^2$ is $a\pm b$.

Hence *a trinomial is a perfect square when two of its terms are squares, and the third is the double product of the roots of these squares.*

Whenever, therefore, we meet with a quantity of this description, we may know that its square root is a binomial; and the root may be found *by extracting the roots of the two terms which are complete squares, and connecting them by the sign of the other term.*

Ex. 1. Find the square root of $a^2+4ab+4b^2$.

The two terms, a^2 and $4b^2$ are complete squares, and the third term $4ab$ is twice the product of the roots a and $2b$; hence $a+2b$ is the root required.

Ex. 2. Find the square root of $9a^2-24ab+16b^2$.

Ex. 3. Find the square root of $9a^4-30a^3b+25a^2b^2$.

Ex. 4. Find the square root of $4a^2+14ab+9b^2$.

(153.) *No binomial can be a perfect square.* For the square of a monomial is a monomial; and the square of a binomial consists of three distinct terms, which do not admit of being reduced with each other.

Thus such an expression as

$$a^2+b^2$$

is not a square; it wants the term $\pm 2ab$ to render it the square of $a\pm b$. This remark should be continually borne in mind as beginners often put the square root of a^2+b^2 equal to $a+b$.

IRRATIONAL QUANTITIES, OR SURDS.

(154.) A *rational quantity* is one which can be expressed in finite terms, and without any radical sign; as a, $5a^2$, &c.

Irrational quantities, or surds, are quantities affected with a radical sign, and which have no exact root, or a root which can be exactly expressed in numbers.

Thus, $\sqrt{3}$ is a surd, because the square root of 3 can not be expressed in numbers with perfect exactness.

In decimals it is 1.7320508 nearly.

(155.) We have seen, *Art.* 144, that in order to extract the square root of a monomial, we must divide each of its exponents by 2.

Thus the square root of a^2 is a^1 or a; that of a^4 is a^2; that of a^6 is a^3, and so on; and as this principle is general, the square root of a^3 must necessarily be $a^{\frac{3}{2}}$, and that of a^5 must be $a^{\frac{5}{2}}$; and, in the same manner, we shall have $a^{\frac{1}{2}}$ for the square root of a^1. Whence we see that

$a^{\frac{1}{2}}$ is equal to $\sqrt{a}$,

$a^{\frac{3}{2}}$ is the same as $\sqrt{a^3}$,

$a^{\frac{n}{2}}$ is equivalent to $\sqrt{a^n}$,

&c., &c.

We have also seen, *Art.* 149, that in order to extract any root of a monomial, we must divide the exponent of each letter by the index of the required root.

Thus, the cube root of a^3 is a^1, or a; the cube root of a^6 is a^2; the cube root of a^9 is a^3, and so on. So, also, the cube root of a^2 is $a^{\frac{2}{3}}$; the cube root of a^4 is $a^{\frac{4}{3}}$; the cube root of a, or a^1, is $a^{\frac{1}{3}}$. Whence it appears that

$a^{\frac{1}{3}}$ is the same as $\sqrt[3]{a}$,

$a^{\frac{4}{3}}$ is equivalent to $\sqrt[3]{a^4}$,

$a^{\frac{n}{3}}$ is equivalent to $\sqrt[3]{a^n}$,
&c., &c.

In the same manner, the fourth root of a is $a^{\frac{1}{4}}$, which expression has therefore the same value as $\sqrt[4]{a}$; the fifth root of a will be $a^{\frac{1}{5}}$, which is, consequently, equivalent to $\sqrt[5]{a}$, and the same principle may be extended to all roots of a higher degree.

(156.) Other fractional exponents are to be understood in the same way. Thus, if we have $a^{\frac{5}{4}}$, this means that we must first take the fifth power of a, and then extract its fourth root; so that $a^{\frac{5}{4}}$ is the same as $\sqrt[4]{a^5}$.

So, also, to find the value of $a^{\frac{m}{n}}$, we must first take the mth power of a, which is a^m, and then extract the nth root of that power: so that $a^{\frac{m}{n}}$ is the same as $\sqrt[n]{a^m}$.

Hence *the numerator of a fractional exponent denotes the power, and the denominator the root to be extracted.*

Again, let it be required to extract the cube root of $\frac{1}{a^4}$.

In the first place, $\frac{1}{a^4}=a^{-4}$. Now, to extract the cube root of a^{-4}, we must divide its exponent by 3, which gives us

$$a^{-\frac{4}{3}}.$$

But the cube root of $\frac{1}{a^4}$ may also be represented by $\frac{1}{a^{\frac{4}{3}}}$.

Hence $\frac{1}{a^{\frac{4}{3}}}$ is equivalent to $a^{-\frac{4}{3}}$.

So, also, $\frac{1}{a^{\frac{1}{2}}}$ is equivalent to $a^{-\frac{1}{2}}$,

$\frac{1}{a^{\frac{1}{n}}}$ is equivalent to $a^{-\frac{1}{n}}$,

$\frac{1}{a^{\frac{m}{n}}}$ is equivalent to $a^{-\frac{m}{n}}$.

Thus we see that the principle of *Art.* 69, that a factor may be transferred from the numerator to the denominator of a fraction, or from the denominator to the numerator by changing the sign of its exponent, is applicable also to *fractional* exponents.

We may therefore entirely reject the radical signs hitherto made use of, and employ, in their stead, the fractional exponents which we have just explained; and, indeed, many of the difficulties in the reduction of radical quantities disappear when fractional exponents are substituted for the radical signs.

PROBLEM I.

To reduce surds to their most simple forms.

(157.) Surds may frequently be simplified by the application of the following principle: *the square root of the product of two or more factors is equal to the product of the square roots of those factors.*

Or, in algebraic language,

$$\sqrt{ab}=\sqrt{a}\times\sqrt{b}.$$

For each member of this equation squared will give the same quantity.

Thus, the square of $\sqrt{ab}$ is ab.

And the square of $\sqrt{a}\times\sqrt{b}$ is $(\sqrt{a})^2\times(\sqrt{b})^2=ab$.

Hence, since the *squares* of the quantities $\sqrt{ab}$ and $\sqrt{a}\times\sqrt{b}$ are equal, the quantities themselves must be equal.

Let it be required to reduce $\sqrt{4a}$ to its most simple form.

This expression may be put under the form $\sqrt{4}\times\sqrt{a}$.

But $\sqrt{4}$ is equal to 2.

Hence, $\sqrt{4a}=\sqrt{4}\times\sqrt{a}=2\sqrt{a}=2a^{\frac{1}{2}}$.

$2\sqrt{a}$ is considered a *simpler* form than $\sqrt{4a}$, for reasons which will be better understood hereafter.

Again, reduce $\sqrt{48}$ to its most simple form.

$\sqrt{48}$ is equal to $\sqrt{16\times3}=\sqrt{16}\times\sqrt{3}=4\sqrt{3}$.

Therefore, in order to simplify a monomial radical of the second degree, *separate it into two factors, one of which is a perfect square; extract its root; and prefix it to the other factor with the radical sign between them.*

In the expressions $2\sqrt{a}$ and $4\sqrt{3}$, the quantities 2 and 4 are called *the coefficients of the radical.*

EXAMPLES.

1. Reduce $2\sqrt{32}$ to its most simple form.

Ans. $8\sqrt{2}$.

2. Reduce $\sqrt{125a^3}$ to its most simple form.

Ans. $5a\sqrt{5a}$.

3. Reduce $\sqrt{98ab^4}$ to its most simple form.

Ans. $7b^2\sqrt{2a}$.

4. Reduce $\sqrt{294ab^2}$ to its most simple form.
5. Reduce $7\sqrt{80abc^3}$ to its most simple form.
6. Reduce $\sqrt{98a^2x^6y^2}$ to its most simple form.
7. Reduce $\sqrt{45a^2b^3c^2d}$ to its most simple form.
8. Reduce $\sqrt{864a^2b^5c^{11}}$ to its most simple form.

(158.) Surds of *any* degree may be simplified by the application of the following principle, which is merely an extension of that already proved in the preceding Article.

The n*th root of the product of any number of factors is equal to the product of the* n*th roots of those factors.*

Or, in algebraic language,

$$\sqrt[n]{ab}=\sqrt[n]{a}\times\sqrt[n]{b}.$$

For, raise each of these expressions to the nth power, and we shall obtain the same result.

Thus, the nth power of $\sqrt[n]{ab}$ is ab.

And the nth power of $\sqrt[n]{a}\times\sqrt[n]{b}$ is $(\sqrt[n]{a})^n\times(\sqrt[n]{b})^n=ab$.

Hence, since the *same* powers of the quantities $\sqrt[n]{ab}$ and $\sqrt[n]{a}\times\sqrt[n]{b}$ are equal, the quantities themselves must be equal.

Let it be required to reduce $\sqrt[3]{8a^2}$ to its most simple form.

This is equivalent to $\sqrt[3]{8}\times\sqrt[3]{a^2}$, which is equal to $2\sqrt[3]{a^2}$.

Again, take the expression

$$\sqrt[4]{48a^6}.$$

This is equivalent to $\sqrt[4]{16a^4}\times\sqrt[4]{3a}$, which is equal to $2a\sqrt[4]{3a}$.

Hence, to simplify a monomial radical of *any degree*, we have the following

RULE.

Separate the quantity into two factors, one of which is an ex

act power of the same name with the root; extract its root; and prefix it to the other factor with the radical sign between them.

In the expressions $2\sqrt[3]{a^2}$ and $2a\sqrt[4]{3a}$, the quantities 2 and $2a$ placed before the radical sign are called the *coefficients of the radical.*

EXAMPLES.

1. Reduce $\sqrt[3]{56a^5b^6}$ to its most simple form.

Ans. $2ab^2\sqrt[3]{7a^2}$.

2. Reduce $\sqrt[3]{54a^4b^3c^2}$ to its most simple form.

Ans. $3ab\sqrt[3]{2ac^2}$.

3. Reduce $\sqrt[4]{48a^5b^8c^6}$ to its most simple form.

Ans. $2ab^2c\sqrt[4]{3ac^2}$.

4. Reduce $\sqrt[5]{192a^7bc^{12}}$ to its most simple form.

5. Reduce $\sqrt[3]{192a^4b^2c^3}$ to its most simple form.

6. Reduce $9\sqrt[3]{81b^2}$ to its most simple form.

(159.) There is another principle which can frequently be employed to advantage in simplifying radicals.

The *square* of the *cube* of a is equal to the *sixth power* of a.

For the square of the cube of a is $a^3\times a^3$, which equals $a^{3+3}=a^6$.

So, also, the *fourth power* of the *cube* of a is equal to the *twelfth power* of a.

For
$$(a^3)^4=a^3\times a^3\times a^3\times a^3$$
$$=a^{3+3+3+3}$$
$$=a^{12}.$$

And, in general, the mth power of the nth power of any quantity is equal to the mnth power of that quantity.

That is $(a^n)^m=a^{mn}$.

Hence, conversely,

The mnth root of any quantity is equal to the mth root of the nth root of that quantity.

Thus, the fourth root = the square root of the square root;

" the sixth root = the square root of the cube root, or the cube root of the square root;

" the eighth root = the square root of the fourth root, or the fourth root of the square root;

" the ninth root = the cube root of the cube root.

Hence, *when the index of a root is the product of two or more factors, we may obtain the root required by extracting in succession the roots denoted by those factors.*

Ex. 1. Let it be required to extract the sixth root of 64.

The sixth root is equal to the cube root of the square root.

The square root of 64 is 8,
and the cube root of 8 is 2.

Hence the sixth root of 64 is 2.

Ex. 2. Let it be required to extract the eighth root of 256.

The eighth root is equal to the fourth root of the square root; or to the square root of the square root of the square root.

The square root of 256 is 16,
and the fourth root of 16 is 2.

Hence the eighth root of 256 is 2.

When one of the roots can be extracted, and the other can not, a radical may be *simplified* by extracting one of the roots.

Thus, the fourth root of 9 is equal to the square root of the square root of 9; that is, the square root of 3.

Or, algebraically, $\sqrt[4]{9}=\sqrt{3}$.

Ex. 3. Reduce $\sqrt[6]{4a^2}$ to its most simple form.

Ans. $\sqrt[3]{2a}$.

Ex. 4. Reduce $\sqrt[4]{36a^2b^2}$ to its most simple form.

Ex. 5. Reduce $\sqrt[mn]{a^n}$ to its most simple form.

Ex. 6. Reduce $\sqrt[8]{25a^4b^2c^6}$ to its most simple form.

PROBLEM II.

(160.) *To reduce a rational quantity to the form of a surd.*

The square root of the square of a is obviously a; that is,

$$a=\sqrt{a^2}=a^{\frac{2}{2}}.$$

So, also, the cube root of the cube of a is a;

that is, $$a=\sqrt[3]{a^3}=a^{\frac{3}{3}}.$$

Hence, to reduce a rational quantity to the form of a surd, we have the following

RULE.

Raise the quantity to a power of the same name with the given root, and then apply the corresponding radical sign.

EXAMPLES.

1. Reduce 3 to the form of the square root.

Here $3\times3=3^2=9$; whence $3=\sqrt{9}$. *Ans.*

2. Reduce ax to the form of the square root.

Ans. $\sqrt{a^2x^2}$, or $(a^2x^2)^{\frac{1}{2}}$.

3. Reduce $2x^2$ to the form of the cube root.

Ans. $\sqrt[3]{8x^6}$.

4. Reduce $5+b$ to the form of the square root.
5. Reduce $-3x$ to the form of the cube root.
6. Reduce $-\frac{1}{2}x^2$ to the form of the fourth root.
7. Reduce a^2b^4 to the form of the square root.
8. Reduce a^m to the form of the nth root.

It will be observed, that this Problem is nearly the reverse of the preceding, and, consequently, brings quantities into a *less simple* form; nevertheless, this form is sometimes better suited to subsequent operations, as will be seen hereafter.

PROBLEM III.

(161.) *To reduce surds which have different indices to others of the same value having a common index.*

Ex. 1. Reduce $a^{\frac{1}{2}}$ and $a^{\frac{1}{3}}$ to surds having the same radical sign.

From the preceding Article, it is obvious that the square root of a is equal to the sixth root of the cube of a;

that is, $$a^{\frac{1}{2}}=a^{\frac{3}{6}}=\sqrt[6]{a^3}.$$

So, also, $$a^{\frac{1}{3}}=a^{\frac{2}{6}}=\sqrt[6]{a^2}.$$

Thus, the quantities $a^{\frac{1}{2}}$ and $a^{\frac{1}{3}}$ are reduced to $\sqrt[6]{a^3}$ and $\sqrt[6]{a^2}$, which are of the same value, and have the common index 6.

Ex. 2. Reduce $3^{\frac{1}{2}}$ and $2^{\frac{1}{3}}$ to a common index.

$$3^{\frac{1}{2}}=3^{\frac{3}{6}}=(3^3)^{\frac{1}{6}}=27^{\frac{1}{6}}.$$

$$2^{\frac{1}{3}}=2^{\frac{2}{6}}=(2^2)^{\frac{1}{6}}=4^{\frac{1}{6}}.$$

Hence $\sqrt[6]{27}$ and $\sqrt[6]{4}$ are the quantities required.

Whence we derive the following

RULE.

Reduce the fractional exponents to a common denominator; raise each quantity to the power denoted by the numerator of its reduced exponent; and take the root denoted by the common denominator.

Ex. 3. Reduce $2^{\frac{1}{3}}$ and $4^{\frac{1}{4}}$ to a common index.

Ans. $\sqrt[6]{4}$ and $\sqrt[6]{8}$.

Ex. 4. Reduce a^2 and $a^{\frac{1}{2}}$ to a common index.

Ex. 5. Reduce $a^{\frac{1}{2}}$ and $b^{\frac{2}{3}}$ to a common index.

Ex. 6. Reduce $5^{\frac{2}{3}}$ and $7^{\frac{3}{4}}$ to a common index.

Ex. 7. Reduce $a^{\frac{1}{n}}$ and $b^{\frac{1}{m}}$ to a common index.

PROBLEM IV.

To add surd quantities together.

(162.) Two radicals are *similar* when they have the same index, and the same quantity under the radical sign.

Thus, $3\sqrt{a}$ and $5\sqrt{a}$ are similar radicals.

So, also, $7\sqrt[3]{b}$ and $10\sqrt[3]{b}$ are similar radicals.

But $\sqrt{a}$ and $\sqrt[3]{a}$ are *not* similar radicals; for, although they have the same quantity under the radical sign, they have not the same index.

Ex. 1. Find the sum of $2\sqrt{a}$ and $3\sqrt{a}$.

As these are *similar* radicals, we may unite their coefficients by the usual rule; for it is evident that twice the square root of a and three times the square root of a make five times the square root of a. Hence the following

RULE.

When the radicals are similar, add the coefficients, and annex the radical part.

But if the quantities are dissimilar, and can not be made similar by the reductions in the preceding Articles, they can only be connected together by the sign of addition.

Ex. 2. Add $\sqrt{6}$ to $2\sqrt{6}$.

Ans. $3\sqrt{6}$.

Ex. 3. Add $5\sqrt[3]{a}$ and $-2\sqrt[3]{a}$.

Ex. 4. Add $a\sqrt{b+c}$ and $x\sqrt{b+c}$.

If the radical parts are *originally different*, they must, if possible, be made alike by the preceding methods.

Ex. 5. Add $\sqrt{27}$ to $\sqrt{48}$.

Here $\sqrt{27}=\sqrt{9\times3}=3\sqrt{3}$,

and $\sqrt{48}=\sqrt{16\times3}=4\sqrt{3}$.

Whence their sum $=7\sqrt{3}$.

Ex. 6. Add together $\sqrt[3]{500}$ and $\sqrt[3]{108}$.

Ans. $8\sqrt[3]{4}$.

Ex. 7. Add together $4\sqrt{147}$ and $3\sqrt{75}$.

Ans. $43\sqrt{3}$.

Ex. 8. Add together $3\sqrt{\frac{2}{5}}$ and $2\sqrt{\frac{1}{10}}$.

Here $3\sqrt{\frac{2}{5}}=3\sqrt{\frac{10}{25}}=\frac{3}{5}\sqrt{10}$,

and $2\sqrt{\frac{1}{10}}=2\sqrt{\frac{10}{100}}=\frac{2}{10}\sqrt{10}$.

Whence their sum $=\frac{4}{5}\sqrt{10}$.

Ex. 9. Add together $\sqrt{72}$ and $\sqrt{128}$.

Ex. 10. Add together $\sqrt{180}$ and $\sqrt{405}$.

Ex. 11. Add together $\sqrt[3]{40}$ and $\sqrt[3]{135}$.

Ex. 12. Add together $8\sqrt[4]{32}$ and $5\sqrt[4]{2}$.

PROBLEM V.

To find the difference of surd quantities.

(163.) It is evident that the subtraction of surd quantities may be performed in the same manner as addition, except that the signs in the subtrahend are to be changed according to *Art.* 43.

Ex. 1. Required to find the difference between $\sqrt{448}$ and $\sqrt{112}$.

Here $\sqrt{448}=\sqrt{64\times7}=8\sqrt{7}$,

and $\sqrt{112}=\sqrt{16\times7}=4\sqrt{7}$.

Whence the difference $=4\sqrt{7}$.

Ex. 2. Find the difference between $\sqrt[3]{192}$ and $\sqrt[3]{24}$.

Here $\sqrt[3]{192}=\sqrt[3]{64\times3}=4\sqrt[3]{3}$,

and $\sqrt[3]{24}=\sqrt[3]{8\times3}=2\sqrt[3]{3}$.

Whence the difference $=2\sqrt[3]{3}$.

Ex. 3. Find the difference between $5\sqrt{20}$ and $3\sqrt{45}$.

Here $5\sqrt{20}=5\sqrt{4\times5}=10\sqrt{5}$,

and $3\sqrt{45}=3\sqrt{9\times5}=\ 9\sqrt{5}$.

Whence the difference $=\ \sqrt{5}$.

Ex. 4. Find the difference between $2\sqrt{50}$ and $\sqrt{18}$.

Ex. 5. Find the difference between $2\sqrt[3]{320}$ and $3\sqrt[3]{40}$.

Ex. 6. Find the difference between $\sqrt{80a^4x}$ and $\sqrt{20a^2x^3}$.

Ex. 7. Find the difference between $2\sqrt{72a^2}$ and $\sqrt{162a^2}$.

PROBLEM VI.

To multiply surd quantities together.

(164.) Let it be required to multiply $\sqrt[n]{a}$ by $\sqrt[n]{b}$.

The product will be $\sqrt[n]{ab}$.

For if we raise each of these quantities to the power of n, we obtain the same result, ab; hence these two expressions are equal. We therefore have the following

RULE.

When the surds have the same index, multiply the quantities under the sign by each other, and prefix the common radical sign. If there are coefficients, these must be multiplied separately.

Ex. 1. Required the product of $3\sqrt{8}$ and $2\sqrt{6}$.

Here $3\sqrt{8}$,

multiplied by $2\sqrt{6}$,

gives $6\sqrt{48}=6\sqrt{16\times3}=24\sqrt{3}$. *Ans*

Proof. Square $3\sqrt{8}$, and we obtain $9\times8=72$.

Square $2\sqrt{6}$, and we obtain $4\times6=24$.

72 multiplied by $24=1728$.

Also, $24\sqrt{3}$ squared $=576\times3=1728$.

Ex. 2. Required the product of $5\sqrt{8}$ and $3\sqrt{5}$.

Ans. $30\sqrt{10}$.

Ex. 3. Required the product of $7\sqrt[3]{18}$ and $5\sqrt[3]{4}$.

Ans. $70\sqrt[3]{9}$

Ex. 4. Required the product of $\frac{1}{4}\sqrt[3]{6}$ and $\frac{2}{15}\sqrt[3]{17}$.

Ex. 5. Required the product of $\frac{1}{2}\sqrt[3]{18}$ and $5\sqrt[3]{20}$.

In the preceding examples, let all the results be reduced to their simplest form.

If the surds have *not* the same index, they must first be reduced to a common index, by *Art.* 161.

Ex. 6. Required the product of $\sqrt{2}$ and $\sqrt[3]{3}$.

Here $\sqrt{2}=2^{\frac{3}{6}}=(2^3)^{\frac{1}{6}}=\sqrt[6]{8}$,

and $\sqrt[3]{3}=3^{\frac{2}{6}}=(3^2)^{\frac{1}{6}}=\sqrt[6]{9}$.

Whence the product $=\sqrt[6]{72}$.

(165.) We have seen, in *Art.* 50, that *powers* of the same quantity may be multiplied by adding their exponents. The same principle may be extended to *roots* of the same quantity.

Let it be required to multiply $\sqrt{a}$ by $\sqrt[3]{a}$, or $a^{\frac{1}{2}}$ by $a^{\frac{1}{3}}$.

We have seen, in *Art.* 161, that $a^{\frac{1}{2}}=a^{\frac{3}{6}}$, and $a^{\frac{1}{3}}=a^{\frac{2}{6}}$.

But $a^{\frac{3}{6}}=a^{\frac{1}{6}}\times a^{\frac{1}{6}}\times a^{\frac{1}{6}}$,

and $a^{\frac{2}{6}}=a^{\frac{1}{6}}\times a^{\frac{1}{6}}$.

The product, therefore, is $a^{\frac{1}{6}}\times a^{\frac{1}{6}}\times a^{\frac{1}{6}}\times a^{\frac{1}{6}}\times a^{\frac{1}{6}}=a^{\frac{5}{6}}$.

Hence, *roots of the same quantity may be multiplied by adding their fractional exponents.*

Ex. 1. Multiply $5a^{\frac{1}{2}}$ by $3a^{\frac{1}{3}}$.

Ans. $15a^{\frac{5}{6}}$.

Ex. 2. Multiply $3a^{\frac{2}{3}}$ by $21a^{\frac{1}{4}}$.

Ex. 3. Multiply $3x^{\frac{1}{2}}y^{\frac{1}{3}}$ by $4x^{\frac{1}{2}}y^{\frac{2}{3}}$.

Ex. 4. Multiply $(a+b)^{\frac{1}{n}}$ by $(a+b)^{\frac{1}{m}}$.

(166.) If the rational quantities, instead of being *coefficients* of the radical quantities, are connected with them by the signs + or −, each term of the multiplier must be multiplied into each term of the multiplicand.

1. Let it be required to multiply $3+\sqrt{5}$

by $2-\sqrt{5}$

$6+2\sqrt{5}$

$\quad -3\sqrt{5}-5.$

We obtain the product $6-\sqrt{5}-5$,

which reduces to $1-\sqrt{5}$.

2. Multiply $7+2\sqrt{6}$ by $9-5\sqrt{6}$.

Ans. $3-17\sqrt{6}$.

3. Multiply $9+2\sqrt{10}$ by $9-2\sqrt{10}$.

Ans. 41.

PROBLEM VII.

To divide one surd quantity by another.

(167.) Let it be required to divide $\sqrt[6]{a^3}$ by $\sqrt[6]{a^2}$.

The quotient must be a quantity which, multiplied by the divisor, shall produce the dividend; we thus obtain $\sqrt[6]{a}$; for, according to *Art.* 164, $\sqrt[6]{a^2}\times\sqrt[6]{a}=\sqrt[6]{a^3}$;

that is, $$\frac{\sqrt[6]{a^3}}{\sqrt[6]{a^2}}=\sqrt[6]{a}.$$

Hence the following

RULE.

Quantities under the same radical sign may be divided like rational quantities, the quotient being affected with the common radical sign. If there are coefficients, they must be divided separately.

If the radicals have *not* the same index, we must first reduce them to a common index.

EXAMPLES.

1. It is required to divide $8\sqrt{108}$ by $2\sqrt{6}$.

Here $$\frac{8\sqrt{108}}{2\sqrt{6}}=4\sqrt{18}=4\sqrt{9\times2}=12\sqrt{2}. \textit{ Ans.}$$

2. Divide $8\sqrt[3]{512}$ by $4\sqrt[3]{2}$.

Here $$\frac{8\sqrt[3]{512}}{4\sqrt[3]{2}}=2\sqrt[3]{256}=2\sqrt[3]{64\times4}=8\sqrt[3]{4}. \textit{ Ans.}$$

3. Divide $6\sqrt[3]{54}$ by $3\sqrt[3]{2}$.

Ans. 6.

4. Divide $4\sqrt[3]{72}$ by $2\sqrt[3]{18}$.

5. Divide $4\sqrt{6a^2y}$ by $2\sqrt{3y}$.

Ans. $2a\sqrt{2}$.

6. Divide $16(a^3b)^{\frac{1}{m}}$ by $8(ac)^{\frac{1}{m}}$.

7. Divide $4\sqrt[3]{12}$ by $2\sqrt{3}$.

As the radicals in this last example have not the same index, they must be reduced to a common index.

$$4\sqrt[3]{12}=4(12)^{\frac{1}{3}}=4(12)^{\frac{2}{6}}=4(144)^{\frac{1}{6}}.$$

$$2\sqrt{3}\ =2(3)^{\frac{1}{2}}\ =2(3)^{\frac{3}{6}}\ =2(27)^{\frac{1}{6}}.$$

Hence $$\frac{4(144)^{\frac{1}{6}}}{2(27)^{\frac{1}{6}}}=2(\tfrac{144}{27})^{\frac{1}{6}}=2(\tfrac{16}{3})^{\frac{1}{6}}=2\sqrt[6]{\tfrac{16}{3}}.$$

(168.) We have seen, in *Art.* 67, that, in order to divide quantities expressed by the same letter, we must subtract the exponent of the divisor from the exponent of the dividend. The same principle may be extended to fractional exponents.

Thus, let it be required to divide $a^{\frac{1}{2}}$ by $a^{\frac{1}{3}}$.

According to the preceding Article,

$$\frac{a^{\frac{1}{2}}}{a^{\frac{1}{3}}}=\frac{a^{\frac{3}{6}}}{a^{\frac{2}{6}}}=\sqrt[6]{\frac{a^3}{a^2}}=\sqrt[6]{a}=a^{\frac{1}{6}}.$$

Hence *a root is divided by another root of the same letter or quantity, by subtracting the exponent of the divisor from that of the dividend.*

Ex. 1. Divide $(ab)^{\frac{2}{3}}$ by $(ab)^{\frac{1}{3}}$.

Ans. $(ab)^{\frac{1}{3}}$.

Ex. 2. Divide a^3 by $a^{\frac{1}{2}}$.

Ex. 3. Divide $a^{\frac{2}{3}}$ by $a^{\frac{1}{4}}$.

Ex. 4. Divide $a^{\frac{1}{n}}$ by $a^{\frac{1}{m}}$.

Ex. 5. Divide $4\sqrt{ab}$ by $2\sqrt[3]{ab}$.

Ans. $2\sqrt[6]{ab}$.

PROBLEM VIII.

(169.) *To raise surd quantities to any power.*

Let it be required to find the square of $a^{\frac{1}{3}}$.

The square of a quantity is found by multiplying it by itself once.

Hence the square of $a^{\frac{1}{3}}$ is equal to $a^{\frac{1}{3}}\times a^{\frac{1}{3}}=a^{\frac{1}{3}+\frac{1}{3}}=a^{\frac{2}{3}}$.

That is, $$\left(a^{\frac{1}{3}}\right)^2=a^{\frac{2}{3}}.$$

Again, let it be required to find the cube of $a^{\frac{1}{5}}$.

The cube of a quantity is found by multiplying it by itself twice.

Hence the cube of $a^{\frac{1}{5}}$ is equal to $a^{\frac{1}{5}} \times a^{\frac{1}{5}} \times a^{\frac{1}{5}} = a^{\frac{3}{5}}$;

that is, $\left(a^{\frac{1}{5}}\right)^3 = a^{\frac{3}{5}}$.

In the same manner we should find the nth power of $a^{\frac{1}{m}} = a^{\frac{n}{m}}$.

Hence we have the following

RULE.

Radical quantities are involved by multiplying their fractional exponents by the exponent of the required power.

Ex. 1. Required the fourth power of $\frac{2}{3}a^{\frac{1}{3}}$.

Ex. 2. Required the cube of $\frac{2}{3}\sqrt{3}$.

Ans. $\frac{8}{9}\sqrt{3}$.

Ex. 3. Required the square of $3\sqrt[3]{3}$.

Ex. 4. Required the cube of $17\sqrt{21}$.

Ex. 5. Required the fourth power of $\frac{1}{6}\sqrt{6}$.

Ans. $\frac{1}{36}$.

(170.) *If the radical quantities are connected with others by the signs + and −, they must be involved by a multiplication of the several terms.*

Ex. 1. Required the square of $3+\sqrt{5}$.

$$\begin{array}{r} 3+\sqrt{5} \\ \underline{3+\sqrt{5}} \\ 9+3\sqrt{5} \\ \underline{3\sqrt{5}+5} \\ 9+6\sqrt{5}+5 \end{array}$$

The square is $9+6\sqrt{5}+5$

or $14+6\sqrt{5}$. *Ans.*

Ex. 2. Required the square of $3+2\sqrt{5}$.

These two examples are comprehended under the rule in *Art.* 60, that the square of the sum of two quantities is equal to the square of the first, plus twice the product of the first by the second, plus the square of the second.

Ex. 3. Required the cube of $\sqrt{x}+3\sqrt{y}$.

Ex. 4. Required the fourth power of $\sqrt{3}-\sqrt{2}$.

Ans. $49-20\sqrt{6}$.

PROBLEM IX.

To find the roots of surd quantities.

(171.) A root of a quantity is a factor which, multiplied by itself a certain number of times, will produce the given quantity. But we have seen that a radical quantity is involved by multiplying its exponent by the exponent of the required power. Hence,

To find the roots of surd quantities,

Divide the fractional exponent by the index of the required root.

Thus, the square root of $a^{\frac{1}{3}}$ is $a^{\frac{1}{3}\div 2}=a^{\frac{1}{6}}$.

For, by *Art.* 169, we obtain the square of $a^{\frac{1}{6}}$ by multiplying the exponent $\frac{1}{6}$ by 2;

that is, $$(a^{\frac{1}{6}})^2=a^{\frac{2}{6}}=a^{\frac{1}{3}}.$$

EXAMPLES.

1. Find the square root of $9(3)^{\frac{1}{3}}$.

Here $\left(9(3)^{\frac{1}{3}}\right)^{\frac{1}{2}}=9^{\frac{1}{2}}\times 3^{\frac{1}{3}\div 2}=3(3)^{\frac{1}{6}}=3\sqrt[6]{3}$. *Ans.*

2. Required the cube root of $\frac{1}{8}\sqrt{2}$.

Ans. $\frac{1}{2}\sqrt[6]{2}$

3. Required the square root of 10^3.
4. Required the cube root of $\frac{8}{27}a^4$.
5. Required the fourth root of $\frac{16}{81}a^{\frac{2}{3}}$.
6. Required the cube root of $\frac{64}{125}a^6$.
7. Required the cube root of $\frac{a}{3}\sqrt{\frac{a}{3}}$.

Ans. $\sqrt{\frac{a}{3}}$.

PROBLEM X.

To find multipliers which shall cause surds to become rational.

(172.) I. When the surd is a *monomial.*

The quantity $\sqrt{a}$ is rendered rational by multiplying it by $\sqrt{a}$.

For $$\sqrt{a}\times\sqrt{a}=a^{\frac{1}{2}}\times a^{\frac{1}{2}}=a.$$

So, also, $a^{\frac{1}{3}}$ is rendered rational by multiplying it by $a^{\frac{2}{3}}$.

For $$a^{\frac{1}{3}}\times a^{\frac{2}{3}}=a^{\frac{3}{3}}=a.$$

Also, $a^{\frac{1}{4}}$ is rendered rational by multiplying it by $a^{\frac{3}{4}}$.

For $$a^{\frac{1}{4}}\times a^{\frac{3}{4}}=a^{\frac{4}{4}}=a.$$

In general, $a^{\frac{1}{n}}$ is rendered rational by multiplying it by $a^{\frac{n-1}{n}}$.

For $$a^{\frac{1}{n}}\times a^{\frac{n-1}{n}}=a^{\frac{n-1+1}{n}}=a^{\frac{n}{n}}=a.$$

Hence we deduce the following

RULE.

Multiply the surd by the same quantity having such an exponent as, when added to the exponent of the given surd, shall be equal to unity.

(173.) II. When the surd is a *binomial.*

If the binomial contains only the *square root, multiply the given binomial by the same expression with the sign of one of its terms changed, and it will give a rational product.*

Ex. 1. The expression $\sqrt{a}+\sqrt{b}$

Multiplied by $\sqrt{a}-\sqrt{b}$

$$a+\sqrt{ab}$$

$$-\sqrt{ab}-b$$

Gives a product $a \quad -b$, which is rational.

Ex. 2. Find a multiplier which shall render $5+\sqrt{3}$ rational

Given surd, $5+\sqrt{3}$

Multiplier, $5-\sqrt{3}$

Product, $25-3=22$, as required.

These two examples are comprehended under the Rule in *Art.* 62, the product of the sum and difference of two quantities is equal to the difference of their squares.

Ex. 3. Find a multiplier that shall make $\sqrt{5}+\sqrt{3}$ rational, and determine the product.

Ex. 4. Find a multiplier that shall make $\sqrt{5}-\sqrt{x}$ rational and determine the product.

Ex. 5. Find a multiplier that shall make $\sqrt{a}-\sqrt{abc}$ rational.

III. When the surd is a *trinomial*, it may be reduced, by successive multiplications, first to a binomial surd, and then to a rational quantity.

Ex. 1. Find multipliers that shall make $\sqrt{5}+\sqrt{3}-\sqrt{2}$ rational.

Given surd, $\sqrt{5}+\sqrt{3}-\sqrt{2}$
First multiplier, $\sqrt{5}+\sqrt{3}+\sqrt{2}$

$$5+\sqrt{15}-\sqrt{10}$$
$$+\sqrt{15}+3-\sqrt{6}$$
$$+\sqrt{10}+\sqrt{6}-2$$

First product, $2\sqrt{15}+6$
Second multiplier, $2\sqrt{15}-6$

$$60+12\sqrt{15}$$
$$-12\sqrt{15}-36$$

Second product, $60-36=24$, a rational quantity.

Ex. 2. Find multipliers that shall make $\sqrt{a}+\sqrt{b}+\sqrt{c}$ rational, and determine the product.

PROBLEM XI.

(174.) *To reduce a fraction containing surds to another having a rational numerator or denominator.*

RULE.

Multiply both numerator and denominator by a factor which will render either of them rational, as the case may require.

Ex. 1. If both terms of the fraction $\frac{\sqrt{a}}{\sqrt{b}}$ be multiplied by $\sqrt{a}$, it will become $\frac{a}{\sqrt{ab}}$, in which the *numerator* is rational. Or if both terms be multiplied by $\sqrt{b}$, it will become $\frac{\sqrt{ab}}{b}$, in which the *denominator* is rational.

Ex. 2. Reduce the fraction $\frac{2}{\sqrt{3}}$ to one that shall have a rational denominator.

Ans. $\frac{2\sqrt{3}}{3}$.

Ex. 3. Reduce $\frac{1}{\sqrt{5}-\sqrt{2}}$ to a fraction having a rational denominator.

Ans. $\frac{\sqrt{5}+\sqrt{2}}{3}$

Ex. 4. Reduce $\frac{\sqrt{2}}{3-\sqrt{2}}$ to a fraction having a rational denominator.

Ans. $\frac{3\sqrt{2}+2}{7}$.

Ex. 5. Reduce $\frac{a-\sqrt{b}}{a+\sqrt{b}}$ to *a* fraction having a rational denominator.

Ex. 6. Reduce $\frac{4}{\sqrt{3}+\sqrt{2}+1}$ to an expression having a rational denominator.

Ans. $2+\sqrt{2}-\sqrt{6}$.

Ex. 7. Reduce $\sqrt{5}+\sqrt{2}$ to a fraction having a rational numerator.

(175.) The *utility* of the preceding transformations may be illustrated by computing the *numerical value* of a fractional surd.

Ex. 1. Suppose it is required to find the square root of $\frac{3}{7}$; that is, it is required to find the value of the fraction $\frac{\sqrt{3}}{\sqrt{7}}$.

If we make the denominator rational, we shall have $\frac{\sqrt{21}}{7}$, in which it is only necessary to extract the square root of the numerator, and the value of the fraction is found to be 0.6546.

Ex. 2. It is required to find the value of the fraction $\frac{7\sqrt{5}}{\sqrt{11}+\sqrt{3}}$.

Making the denominator rational, we have $\frac{7\sqrt{55}-7\sqrt{15}}{8}$, the value of which is 3.1003.

Ex. 3. Required the value of the expression $\frac{\sqrt{6}}{\sqrt{7}+\sqrt{3}}$.

Ans. 0.5595.

Ex. 4. Required the value of the expression

$$\frac{\sqrt{3}}{2\sqrt{8}+3\sqrt{5}-7\sqrt{2}}.$$

Ans. 0.7025.

Ex. 5. Required the value of the expression $\frac{9+2\sqrt{10}}{9-2\sqrt{10}}$.

Ans. 5.7278.

PROBLEM XII.

(176.) *To free an equation from radical quantities.*

This may generally be done by successive involutions. For this purpose, we first free the equation from fractions. If there is but *one* radical expression, we bring that to stand alone on one side of the equation, and involve the whole equation to a power denoted by the index of the radical.

Ex. 1. Free the equation

$$\frac{a+\sqrt{2ax+x^2}}{a}=b$$

from radical quantities.

Clearing of fractions, and transposing a, we obtain

$$\sqrt{2ax+x^2}=ab-a.$$

The square of this equation is

$$2ax+x^2=a^2b^2-2a^2b+a^2,$$

which is free from radical quantities.

Ex. 2. Free the equation

$$x+\sqrt{a^2+x^2}=\frac{2a^2}{\sqrt{a^2+x^2}}$$

from radical quantities.

If the equation contains *two* radical expressions, combined with other terms which are rational, it will generally be best to bring one of the radicals to stand alone on one side of the equation before involution. One of the radicals will thus be made to disappear, and, by repeating the operation, the remaining radical may be exterminated.

Ex. 3. Free the equation

$$\sqrt{a+x}+\sqrt{b+y}=c$$

from radical quantities.

Transposing one of the radicals, we obtain

$$\sqrt{a+x}=c-\sqrt{b+y}.$$

Squaring, we have

$$a+x=c^2-2c\sqrt{b+y}+b+y.$$

Transposing, so as to bring the radical to stand alone, we have

$$2c\sqrt{b+y}=c^2+b+y-a-x,$$

which may be freed from radicals by squaring a second time.

Sometimes the two radicals may be of such a form that it is best to bring both to the same member of the equation before involution.

When an equation contains *several* radical quantities, it may generally be freed from them by successive involutions, but the best mode of procedure can only be determined by trial.

Ex. 4. Free the equation

$$\sqrt{2x+7}+\sqrt{3x-18}=\sqrt{7x+1}$$

from radical quantities.

Ans. $6x^2-15x-126=x^2+12x+36.$

When an equation contains a fraction involving radical quantities in both numerator and denominator, it is sometimes best to render the denominator rational by Problem XI.

Ex. 5. Free the equation

$$\frac{\sqrt{x}+\sqrt{x-a}}{\sqrt{x}-\sqrt{x-a}}=\frac{an^2}{x-a}$$

from radical quantities.

Multiply both terms of the first fraction by $\sqrt{x}+\sqrt{x-a}$, and we have

$$\frac{(\sqrt{x}+\sqrt{x-a})^2}{x-(x-a)}=\frac{an^2}{x-a},$$

or

$$(\sqrt{x}+\sqrt{x-a})^2=\frac{a^2n^2}{x-a}.$$

Extracting the square root, we obtain

$$\sqrt{x}+\sqrt{x-a}=\frac{an}{\sqrt{x-a}}.$$

Clearing of fractions, we have

$$\sqrt{x^2-ax}+x-a=an,$$

which is easily freed from radicals.

Ex. 6. Free the equation

$$\frac{x+\sqrt{x}}{x-\sqrt{x}}=\frac{x^2-x}{4}$$

from radical quantities.

Ans. $x^2\mp 4x+4=x$

Ex. 7. Free the equation

$$\frac{x-\sqrt{x+1}}{x+\sqrt{x+1}}=\frac{5}{11}$$

from radical quantities.

Ans. $9x^2=64x+64$.

Ex. 8. Free the equation

$$\frac{\sqrt{a^2-x^2}-\sqrt{b^2+x^2}}{\sqrt{a^2-x^2}+\sqrt{b^2+x^2}}=\frac{m}{n}$$

from radical quantities.

(177.) The preceding rules for the reduction of radicals, are exact so long as we treat of *absolute numbers*, but require some modifications when we consider *imaginary expressions*, such as $\sqrt{-3}$, $\sqrt{-a}$, &c.

Let it be required, for example, to determine the product of $\sqrt{-a}$ by $\sqrt{-a}$.

By the rule given in *Art.* 164,

$$\begin{aligned}\sqrt{-a}\times\sqrt{-a}&=\sqrt{-a\times -a}\\ &=\sqrt{+a^2}.\end{aligned}$$

Now, $\sqrt{+a^2}=\pm a$, so that there is apparently a doubt as to the sign with which a ought to be affected in order to answer the question. However, the true result is $-a$, because any quantity must be equal to the square of its square root.

That is, $\sqrt{-a}\times\sqrt{-a}$ is the same as $(\sqrt{-a})^2$, and, consequently, is equal to $-a$.

Again, let it be required to determine the product of $\sqrt{-a}$ by $\sqrt{-b}$.

By the rule in *Art.* 164.

$$\begin{aligned}\sqrt{-a}\times\sqrt{-b}&=\sqrt{-a\times -b}\\ &=\sqrt{+ab}\\ &=\pm\sqrt{ab}.\end{aligned}$$

The result, however, is not properly ambiguous, and should be $-\sqrt{ab}$; for we have, according to *Art.* 148,

$$\sqrt{-a} = \sqrt{a}.\sqrt{-1},$$

and
$$\sqrt{-b} = \sqrt{b}.\sqrt{-1}.$$

Hence

$$\sqrt{-a} \times \sqrt{-b} = \sqrt{ab}(\sqrt{-1})^2$$
$$= \sqrt{ab} \times -1$$
$$= -\sqrt{ab}.$$

In the same manner we shall find for the different powers of $\sqrt{-1}$ the following results.

$$\sqrt{-1} = \sqrt{-1}, \text{ the first power.}$$
$$(\sqrt{-1})^2 = -1, \text{ the second power.}$$
$$(\sqrt{-1})^3 = -1 \times \sqrt{-1}$$
$$= -\sqrt{-1}, \text{ the third power.}$$
$$(\sqrt{-1})^4 = (\sqrt{-1})^2 \times (\sqrt{-1})^2$$
$$= -1 \times -1$$
$$= +1, \text{ the fourth power.}$$

Since the four following powers will be found by multiplying -1 by the first, the second, the third, and the fourth powers, we shall again find for the four next powers

$$+\sqrt{-1},\ -1,\ -\sqrt{-1},\ +1;$$

so that all the powers of $\sqrt{-1}$ will form a repeating cycle of these four terms.

Whenever the student is at a loss to determine the product of two imaginary quantities, it is best to resolve each of them into two factors, one the square root of a positive quantity, and the other $\sqrt{-1}$, *Art.* 148.

EXAMPLES.

1. Let it be required to multiply $\sqrt{-9}$ by $\sqrt{-4}$.

Here we have
$$\sqrt{-9} = 3\sqrt{-1},$$

and
$$\sqrt{-4} = 2\sqrt{-1}.$$

Therefore,
$$\sqrt{-9} \times \sqrt{-4} = 3\sqrt{-1} \times 2\sqrt{-1}$$
$$= 6\sqrt{(-1)^2}$$
$$= -6.$$

2. Multiply $1+\sqrt{-1}$ by $1-\sqrt{-1}$.

Ans. 2.

3. Multiply $\sqrt{18}$ by $\sqrt{-2}$.

4. Multiply $5+2\sqrt{-3}$ by $2-\sqrt{-3}$.

SECTION XII.

EQUATIONS OF THE SECOND DEGREE.

(178.) According to *Art.* 96, *quadratic equations, or equations of the second degree, are those in which the highest power of the unknown quantity is a square.*

Quadratic equations are divided into *two classes.*

I. Equations which involve only the square of the unknown quantity and known terms. These are called *pure quadratics.* Of this description are the equations

$$ax^2=b;\ 3x^2+12=150-x^2, \&c.$$

They are sometimes called *quadratic equations of two terms,* because, by transposition and reduction, they can always be exhibited under the general form

$$ax^2=b.$$

II. Equations which involve both the square and the first power of the unknown quantity, together with a known term. These are called *affected* or *complete quadratics.* Of this description are the equations

$$ax^2+bx=c;\ x^2-10x=7;\ \frac{5x^2}{6}-\frac{x}{2}+\frac{3}{4}=8.$$

They are sometimes called quadratic equations of *three terms,* because, by transposition and reduction, they can always be exhibited under the general form

$$ax^2+bx=c.$$

PURE QUADRATIC EQUATIONS.

(179.) The equation

$$ax^2=b$$

is easily solved. Dividing each member by a, it becomes

$$x^2=\frac{b}{a}.$$

Whence $$x=\pm\sqrt{\frac{b}{a}}.$$

If $\frac{b}{a}$ be a particular number, either integral or fractional, we can extract its square root either exactly or approximately by the rules of arithmetic.

It is to be remarked, that since the square both of $+m$ and $-m$ is $+m^2$, so, in like manner, the square of $+\sqrt{\frac{b}{a}}$ and that of $-\sqrt{\frac{b}{a}}$ are both $+\frac{b}{a}$. Hence the above equation is susceptible of *two* solutions, or has *two roots;* that is, there are two quantities which, when substituted for x in the original equation, will render the two members identical. These are

$$+\sqrt{\frac{b}{a}} \text{ and } -\sqrt{\frac{b}{a}}.$$

For, substituting each of these values in the original equation $ax^2=b$, it becomes

$$a\times\left(+\sqrt{\frac{b}{a}}\right)^2=b, \text{ or } a\times\frac{b}{a}=b; \text{ i. e., } b=b,$$

and $$a\times\left(-\sqrt{\frac{b}{a}}\right)^2=b, \text{ or } a\times\frac{b}{a}=b; \text{ i. e., } b=b.$$

EXAMPLE I.

Find the values of x which satisfy the equation

$$4x^2-7=3x^2+9.$$

Transposing terms, $4x^2-3x^2=9+7.$

Reducing, $x^2=16.$

Extracting the square root,

$$x=\pm\sqrt{16}=\pm4.$$

Hence the two values of x are $+4$ and -4, and they may both be verified by substitution in the original equation.

Thus, taking the first value, we have

$$4\times(+4)^2-7=3\times(+4)^2+9,$$

or $$4\times16-7=3\times16+9;$$

that is, $$57=57.$$

Taking the second value of x, we have

$$4\times(-4)^2-7=3\times(-4)^2+9,$$

or $$4\times16-7=3\times16+9, \text{ as before.}$$

From the preceding examples we deduce the following

RULE.

Reduce the equation to the form $ax^2=b$; *then divide by the coefficient of* x^2, *and extract the square root of both members of the equation.*

Ex. 2. Given $x^2-17=130-2x^2$, to find the values of x.

By transposition, $$3x^2=147;$$

therefore, $$x^2=49,$$

and $$x=\pm7.$$

Ex. 3. Given $x^2+ab=5x^2$, to find the values of x.

By transposition, $$ab=4x^2;$$

therefore, $$\pm\sqrt{ab}=2x,$$

and $$x=\frac{\pm\sqrt{ab}}{2}.$$

Ex. 4. Given $x+\sqrt{a^2+x^2}=\frac{2a^2}{\sqrt{a^2+x^2}}$, to find the values of x.

Clearing of fractions, we obtain $x\sqrt{a^2+x^2}+a^2+x^2=2a^2$.

By transposition, $x\sqrt{a^2+x^2}=a^2-x^2$.

Squaring both sides, $a^2x^2+x^4=a^4-2a^2x^2+x^4$;

therefore, $$3a^2x^2=a^4,$$

and $$3x^2=a^2;$$

whence $$x^2=\frac{a^2}{3};$$

therefore, $$x=\pm\frac{a}{\sqrt{3}}.$$

Ex. 5. Given $\frac{4x^2+5}{9}=45$, to find the values of x.

Ex. 6. Given $ax^2-5c=bx^2-3c+d$, to find the values of x.

Ex. 7. Given $\frac{x^2}{3}-3+\frac{5}{12}x^2=\frac{7}{24}-x^2+\frac{299}{24}$, to find the values of x.

Ex. 8. Given $\frac{x^2+3x-7}{x+2+\frac{18}{x}}=1$, to find the values of x.

Ans. $x=\pm 3$.

Clearing of fractions and transposing, we find in each member of this equation a binomial factor, which being canceled, the equation is easily solved.

PROBLEMS.

Prob. 1. What two numbers are those whose sum is to the greater as 10 to 7; and whose sum, multiplied by the less, produces 270?

Let $10x=$ their sum.
Then $7x=$ the greater number,
and $3x=$ the less.
Whence $30x^2=270$,
and $x^2=9$;
therefore, $x=\pm 3$,
and the numbers are ± 21 and ± 9.

Prob. 2. What two numbers are those whose sum is to the greater as m to n; and whose sum, multiplied by the less, is equal to a?

Ans. $\pm\sqrt{\frac{an^2}{m(m-n)}}$ and $\pm\sqrt{\frac{a(m-n)}{m}}$.

Prob. 3. What number is that, the third part of whose square being subtracted from 20, leaves a remainder equal to 8?

Prob. 4. What number is that, the mth part of whose square being subtracted from a, leaves a remainder equal to b?

Ans. $\pm\sqrt{m(a-b)}$

Prob. 5. Find three numbers in the ratio of $\frac{1}{2}$, $\frac{2}{3}$, and $\frac{3}{4}$, the sum of whose squares is 724.

Prob. 6. Find three numbers in the ratio of m, n, and p, the sum of whose squares is equal to a.

Ans.

$$\pm\sqrt{\frac{am^2}{m^2+n^2+p^2}};\ \pm\sqrt{\frac{an^2}{m^2+n^2+p^2}};\ \text{and}\ \pm\sqrt{\frac{ap^2}{m^2+n^2+p^2}}.$$

Prob. 7. Divide the number 49 into two such parts, that the quotient of the greater divided by the less, may be to the quotient of the less divided by the greater, as $\frac{4}{3}$ to $\frac{3}{4}$.

Ans. 21 and 28.

Prob. 8. Divide the number a into two such parts, that the quotient of the greater divided by the less, may be to the quotient of the less divided by the greater, as m to n.

$$\textit{Ans.}\ \frac{a\sqrt{m}}{\sqrt{m}+\sqrt{n}}\ \text{and}\ \frac{a\sqrt{n}}{\sqrt{m}+\sqrt{n}}.$$

Prob. 9. There are two square grass plats, a side of one of which is 10 yards longer than a side of the other, and their areas are as 25 to 9. What are the lengths of the sides?

Prob. 10. There are two squares whose areas are as m to n, and a side of one exceeds a side of the other by a. What are the lengths of the sides?

$$\textit{Ans.}\ \frac{a\sqrt{m}}{\sqrt{m}-\sqrt{n}}\ \text{and}\ \frac{a\sqrt{n}}{\sqrt{m}-\sqrt{n}}.$$

Prob. 11. Two travelers, A and B, set out to meet each other, A leaving Hartford at the same time that B left New York. On meeting, it appeared that A had traveled 18 miles more than B, and that A could have gone B's journey in $15\frac{3}{4}$ hours, but B would have been 28 hours in performing A's journey. What was the distance between Hartford and New York?

Ans. 126 miles.

Prob. 12. From two places at an unknown distance, two bodies, A and B, move toward each other, A going a miles more than B. A would have described B's distance in n hours, and B would have described A's distance in m hours. What was the distance of the two places from each other?

$$\textit{Ans.}\ a\times\frac{\sqrt{m}+\sqrt{n}}{\sqrt{m}-\sqrt{n}}.$$

Prob. 13. A vintner draws a certain quantity of wine out of

a full vessel that holds 256 gallons; and then, filling the vessel with water, draws off the same quantity of liquor as before, and so on for four draughts, when there were only 81 gallons of pure wine left. How much wine did he draw each time?

Ans. 64, 48, 36, and 27 gallons.

Prob. 14. A number a is diminished by the nth part of itself, this remainder is diminished by the nth part of itself, and so on to the fourth remainder, which is equal to b. Required the value of n.

Ans. $\dfrac{\sqrt[4]{a}}{\sqrt[4]{a}-\sqrt[4]{b}}$.

Prob. 15. Two workmen, A and B, were engaged to work for a certain number of days at different rates. At the end of the time, A, who had played 4 of those days, received 75 shillings, but B, who had played 7 of those days, received only 48 shillings. Now had B only played 4 days, and A played 7 days, they would have received the same sum. For how many days were they engaged?

Ans. 19 days.

Prob. 16. A person employed two laborers, allowing them different wages. At the end of a certain number of days, the first, who had played a days, received m shillings, and the second, who had played b days, received n shillings. Now if the second had played a days, and the other b days, they would both have received the same sum. For how many days were they engaged?

Ans. $\dfrac{b\sqrt{m}-a\sqrt{n}}{\sqrt{m}-\sqrt{n}}$ days.

COMPLETE QUADRATIC EQUATIONS.

(180.) Suppose we have the equation

$$x^2-6x+9=1,$$

in which it is required to find the value of x.

Since each member of the equation is a *complete square*, if we extract the square root, we shall obtain a new equation involving only the first power of x, which may be easily solved.

We thus have $x-3=\pm 1$,

and, by transposition,

$$x=3\pm 1=4, \text{ or } 2.$$

In order to verify these values, substitute each of them in place of x in the given equation. Taking the first value, we shall have

$$4^2-6\times 4+9=1\,;$$

that is, $16-24+9=1$, an identical equation.

Taking the second value of x, we obtain

$$2^2-6\times 2+9=1\,;$$

that is, $4-12+9=1$, an identical equation.

Hence we see that a complete quadratic equation is readily solved, provided each member of the equation is a perfect square. But equations seldom occur under this form. Take, for example,

$$x^2-6x=-8.$$

The preceding method seems to be inapplicable, because the first member is not a complete square. We may, however, render it a complete square by the addition of 9, which must also be added to the second member to preserve the equality. The equation thus becomes

$$x^2-6x+9=9-8=1,$$

which is the equation first proposed.

The peculiar difficulty, then, in resolving complete equations of the second degree, consists in rendering the first member an exact square.

(181.) In order to discover a general method of solution, let us take the equation

$$ax^2+bx=c,$$

which is the general form of equations of the second degree. We begin by dividing both members by a, the coefficient of x^2 The equation then becomes

$$x^2+\frac{bx}{a}=\frac{c}{a}.$$

For convenience, let us put $p=\frac{b}{a}$ and $q=\frac{c}{a}$; we shall then have

$$x^2+px=q.$$

We have seen that if we can by any transformation render the first member of this equation the perfect square of a binomial, we can reduce the equation to one of the first degree by extracting the square root.

Now we know that the square of a binomial, $x+a$, or $x^2+2ax+a^2$, is composed of the square of the first term, plus twice the product of the first term by the second, plus the square of the second term.

Hence, considering x^2+px as the first two terms of the square of a binomial, and, consequently, px as being twice the product of the first term of the binomial by the second, it is evident that the second term of this binomial must be $\frac{p}{2}$, for $2\times\frac{p}{2}\times x=px$.

In order, therefore, that the expression x^2+px may be rendered a perfect square, we must add to it the square of this second term $\frac{p}{2}$; that is, the square of half the coefficient of the first power of x; it thus becomes

$$x^2+px+\frac{p^2}{4},$$

which is the square of $x+\frac{p}{2}$. But since we have added $\frac{p^2}{4}$ to the left-hand member of the equation, in order that the equality may not be destroyed, we must add the same quantity to the right-hand member also; the equation thus transformed will become

$$x^2+px+\frac{p^2}{4}=q+\frac{p^2}{4}.$$

Extracting its square root, we have

$$x+\frac{p}{2}=\pm\sqrt{q+\frac{p^2}{4}}.$$

Whence $$x=-\frac{p}{2}\pm\sqrt{q+\frac{p^2}{4}}.$$

We prefix the double sign $\pm$, because the square both of

$+\sqrt{q+\frac{p^2}{4}}$, and also of $-\sqrt{q+\frac{p^2}{4}}$ is $+\left(q+\frac{p^2}{4}\right)$, and every quadratic equation must therefore have two roots.

(182.) From the preceding discussion we deduce the following general

RULE FOR THE SOLUTION OF A COMPLETE QUADRATIC EQUATION.

1. *Transpose all the known quantities to one side of the equation, and all the terms involving the unknown quantity to the other side, and reduce the equation to the form* $ax^2+bx=c$.

2. *Divide each side of the equation by the coefficient of* x^2, *and add to each member the square of half the coefficient of the first power of* x.

3. *Extract the square root of both sides, and the equation will be reduced to one of the first degree, which may be solved in the usual manner.*

EXAMPLE 1.

Solve the equation $x^2-10x=-16$.

Completing the square by adding to each member the square of half the coefficient of the second term, we have

$$x^2-10x+25=25-16=9.$$

Extracting the root, $x-5=\pm 3$.

Hence $x=5\pm 3$.

Ans. $\begin{cases} x=5+3=8, \\ x=5-3=2. \end{cases}$

Thus x has two values, either 8 or 2. To verify them, substitute in the original equation, and we shall have

$$8^2-10\times 8=-16, \text{ i. e., } 64-80=-16;$$

also, $2^2-10\times 2=-16$, *i. e.*, $4-20=-16$,

both of which are identical equations.

EXAMPLE 2.

Solve the equation $x^2+6x=-8$.

Completing the square, $x^2+6x+9=9-8=1$.

Extracting the root, $x+3=\pm 1$.

Hence $x=-3\pm 1$.

Ans. $\begin{cases} x=-3+1=-2, \\ x=-3-1=-4. \end{cases}$

Proof. $(-2)^2+6\times-2=-8$, *i. e.*, $4-12=-8$;
also, $(-4)^2+6\times-4=-8$, *i. e.*, $16-24=-8$.

Hence x has two values, both negative. In verifying them, it is to be observed, that the square of -2 is $+4$, and -2 multiplied by $+6$ gives -12.

EXAMPLE 3.

Solve the equation $x^2+6x=27$.
Completing the square, $x^2+6x+9=27+9=36$.
Extracting the root, $x+3=\pm 6$.
Hence $x=-3\pm 6=+3$, or -9.

EXAMPLE 4.

Solve the equation $x^2-2x=24$.
Here $x=1\pm 5=+6$, or -4.

EXAMPLE 5.

Solve the equation $x^2-8x=-18$.
Completing the square, $x^2-8x+16=16-18=-2$.
Hence $x=4\pm\sqrt{-2}$.
Here both values of x are *imaginary*.

EXAMPLE 6.

Solve the equation $x^2-6x=-9$.
Completing the square, $x^2-6x+9=9-9=0$.
Extracting the root, $x-3=\pm 0$.
Hence $x=3\pm 0$.
Here the two values of x are equal to each other.

Ex. 7. Given $2x^2+8x-20=70$, to find the values of x.
Ans. $x=5$, or -9.

Ex. 8. Given $3x^2-3x+6=5\frac{1}{3}$, to find the values of x.
Ans. $x=\frac{2}{3}$, or $\frac{1}{3}$.

(183.) The Rule given on page 145 for solving a complete quadratic equation is applicable to all cases; nevertheless, a modification of this method is sometimes preferable.

The object is to render the first member of the equation a perfect square. After the equation has been reduced to the form

$$ax^2+bx=c$$

the square may be completed by *multiplying the equation by four times the coefficient of* x^2, *and adding to both sides the square of the coefficient of* x.

Thus the above equation multiplied by $4a$ becomes

$$4a^2x^2+4abx=4ac.$$

Adding b^2 to both members, we have

$$4a^2x^2+4abx+b^2=4ac+b^2.$$

Extracting the square root,

$$2ax+b=\pm\sqrt{4ac+b^2}.$$

Transposing b and dividing by $2a$,

$$x=\frac{-b\pm\sqrt{4ac+b^2}}{2a},$$

which is the same result as would be obtained by the former Rule; but by this new method we have avoided the introduction of fractions in completing the square.

When the coefficient of x^2 is unity, the above Rule becomes, *Multiply the equation by four, and add to each member the square of the coefficient of* x.

Either of these methods of completing the square may be practiced at pleasure; but the first method is to be preferred, except when its application would involve inconvenient fractions.

Ex. 9. Given $\frac{1}{2}x^2-\frac{1}{3}x+20\frac{1}{2}=42\frac{2}{3}$, to find the values of x.

Ans. $x=7$, or $-6\frac{1}{3}$.

Ex. 10. Given $x^2-x-40=170$, to find the values of x.

Ans. $x=15$, or -14

Ex. 11. Given $3x^2+2x-9=76$, to find the values of x.

Ex. 12. Given $\frac{1}{2}x^2-\frac{1}{3}x+7\frac{3}{8}=8$, to find the values of x.

Ex. 13. Given $3x-\frac{6x^2-40}{2x-1}-\frac{3x-10}{9-2x}=2$, to find the values of x.

We must first clear this equation of fractions, which is done by multiplying by the denominators; we thus obtain

$$-12x^3+60x^2-27x+12x^3-54x^2+360-80x-6x^2+23x-10$$
$$=40x-8x^2-18.$$

Here the two terms containing x^3 balance each other, and, uniting similar terms, we obtain

$$8x^2-124x=-368.$$

Dividing by 8, we have

$$x^2-15\tfrac{1}{2}x=-46.$$

Completing the square,

$$x^2-15\tfrac{1}{2}x+\left(\frac{31}{4}\right)^2=-46+\left(\frac{31}{4}\right)^2=\frac{225}{16}.$$

Extracting the root, $x-\frac{31}{4}=\pm\frac{15}{4}$.

Hence $$x=\frac{31}{4}\pm\frac{15}{4}=11\tfrac{1}{2}, \text{ or } 4. \textit{ Ans.}$$

Ex. 14. Given $\frac{90}{x}-\frac{90}{x+1}-\frac{27}{x+2}=0$, to find the values of x.

(184.) The preceding rules will enable us to solve not only quadratic equations, but all equations which can be reduced to the form

$$x^{2n}+px^n=q;$$

that is, *all equations which contain only two powers of the unknown quantity, and in which one of the exponents is double of the other.*

For if, in the above equation, we assume $y=x^n$, then $y^2=x^{2n}$, and it becomes

$$y^2+py=q.$$

Solving this according to the rule, we find

$$x^n=y=-\frac{p}{2}\pm\sqrt{q+\frac{p^2}{4}}.$$

Extracting the nth root of both sides,

$$x=\sqrt[n]{-\frac{p}{2}\pm\sqrt{q+\frac{p^2}{4}}}.$$

EXAMPLE 1.

Given $x^4-25x^2=-144$, to find the values of x.

Assuming $x^2=y$, the above becomes

$$y^2-25y=-144.$$

Whence $y=16$, or 9.

But, since $x^2=y$, $x=\pm\sqrt{y}$.

Therefore, $x=\pm\sqrt{16}$, or $\pm\sqrt{9}$.

Thus x has four values, viz., $+4$, -4, $+3$, -3.

To verify these values:

1st value, $(+4)^4-25\times(+4)^2=-144$, *i. e.*, $256-400=-144$.
2d value, $(-4)^4-25\times(-4)^2=-144$, *i. e.*, $256-400=-144$.
3d value, $(+3)^4-25\times(+3)^2=-144$, *i. e.*, $81-225=-144$.
4th value, $(-3)^4-25\times(-3)^2=-144$, *i. e.*, $81-225=-144$.

EXAMPLE 2.

Given $x^4-7x^2=8$, to find the values of x.

Assuming $x^2=y$, we have

$$y^2-7y=8.$$

Whence $$y=8, \text{ or } -1.$$

Therefore, $x=\pm\sqrt{8}$, or $\pm\sqrt{-1}$, the two last of which roots are *imaginary*.

EXAMPLE 3.

Given $x^6-2x^3=48$, to find the values of x.

Assuming $x^3=y$, the above becomes

$$y^2-2y=48.$$

Whence $$y=8, \text{ or } -6.$$

And since $x^3=y$, therefore $x=\sqrt[3]{y}$.

Hence two of the roots of the above equation are 2 and $-\sqrt[3]{6}$.

This equation has four other roots, which can not be determined by this process.

EXAMPLE 4.

Given $2x-7\sqrt{x}=99$, to find the values of x.

Assuming $\sqrt{x}=y$, this equation becomes

$$2y^2-7y=99.$$

Whence $$y=9, \text{ or } -\frac{11}{2}.$$

And since $\sqrt{x}=y$, therefore $x=y^2$.

Whence $$x=81, \text{ or } \frac{121}{4}.$$

Although the square root of 81 is generally ambiguous, and may be either $+9$ or -9, still, in verifying the preceding values, $\sqrt{x}$ can not be taken equal to -9, because 81 was obtained by multiplying $+9$ by itself. For a like reason, $\sqrt{x}$ can not be taken equal to $+\frac{11}{2}$. A similar remark is applicable to *Exs.* 13 and 14 on the next page, and also to *Ex.* 7, page 186.

EXAMPLE 5.

Given $\sqrt{x+12}+\sqrt[4]{x+12}=6$, to find the values of x.

Assuming $x+12=y$, this equation becomes

$$y^{\frac{1}{2}}+y^{\frac{1}{4}}=6,$$

which evidently belongs to the same class as the previous examples. Completing the square, we shall have

$$y^{\frac{1}{4}}=2, \text{ or } -3.$$

Raising both sides of the equation to the fourth power,

$$y=16, \text{ or } 81.$$

Therefore, $\quad x$ or $(y-12)=4$, or 69.

EXAMPLE 6.

Given $2x^2+\sqrt{2x^2+1}=11$, to find the values of x.

Adding 1 to each member of the equation, it becomes

$$2x^2+1+\sqrt{2x^2+1}=12.$$

Assuming $\quad 2x^2+1=y$, then

$$y+y^{\frac{1}{2}}=12.$$

Completing the square, we find

$$y^{\frac{1}{2}}=3, \text{ or } -4;$$

that is, $\quad \sqrt{2x^2+1}=3$, or -4.

Therefore, $\quad 2x^2+1=9$, or 16,

and $\quad x^2=4$, or $\frac{15}{2}$.

Hence $\quad x=+2,\ -2,\ +\sqrt{\frac{15}{2}},\ -\sqrt{\frac{15}{2}}.$

It may be remarked, that in equations of this kind it is generally unnecessary to substitute a new letter, y, which has been done in the preceding examples simply for the sake of perspicuity.

Ex. 7. Given $x^4+4x^2=12$, to find the values of x.

Ans. $x=\pm\sqrt{2}$, or $\pm\sqrt{-6}$

Ex. 8. Given $x^6-8x^3=513$, to find the values of x.

Ans. $x=3$, or $-\sqrt[3]{19}$.

Ex. 9. Given $x^{\frac{6}{5}}+x^{\frac{3}{5}}=756$, to find the values of x.

Ans. $x=243$, or $-\sqrt[3]{28^5}$.

Ex. 10. Given $\frac{1}{2}x^6-\frac{1}{4}x^3=-\frac{1}{32}$, to find one value of x.

Ans. $x=\frac{1}{2}\sqrt[3]{2}$.

Ex. 11. Given $2x^{\frac{2}{3}}+3x^{\frac{1}{3}}=2$, to find the values of x.

Ans. $x=\frac{1}{8}$, or -8.

Ex. 12. Given $x^4-12x^3+44x^2-48x=9009$, to find the values of x.

This equation may be expressed as follows:

$$(x^2-6x)^2+8(x^2-6x)=9009.$$

Ans. $x=13$, or -7, or $3\pm\sqrt{-90}$.

Ex. 13. Given $\frac{1}{2}x-\frac{1}{3}\sqrt{x}=22\frac{1}{6}$, to find the values of x.

Ans. $x=49$, or $\frac{361}{9}$.

Ex. 14. Given $\sqrt{10+x}-\sqrt[4]{10+x}=2$, to find the values of x

Ans. $x=6$, or -9

Ex. 15. Given $x^6+20x^3-10=59$ to find the values of x.

Ans. $x=\sqrt[3]{3}$, or $\sqrt[3]{-23}$.

Ex. 16. Given $3x^{2n}-2x^n+3=11$, to find the values of x.

Ans. $x=\sqrt[n]{2}$, or $\sqrt[n]{-\frac{4}{3}}$.

Ex. 17. Given $x^2-x\sqrt{3}=x-\frac{1}{2}\sqrt{3}$, to find the values of x

Ans. $\frac{\sqrt{3}+3}{2}$, or $\frac{\sqrt{3}-1}{2}$.

Ex. 18. Given $\sqrt{1+x-x^2}-2(1+x-x^2)=\frac{1}{9}$, to find the values of x.

Ans. $\frac{1}{2}\pm\frac{1}{6}\sqrt{41}$, or $\frac{1}{2}\pm\frac{1}{3}\sqrt{11}$.

(185.) We have seen that every equation of the second degree has *two roots*, or that there are two quantities which, when substituted for x in the original equation, will render the two members identical. In like manner, we shall find that every equation of the third degree has *three roots;* an equation of the fourth degree has *four roots;* and, in general, *an equation has as many roots as it has dimensions.*

Before determining the degree of an equation, it should be freed from fractions, from negative exponents, and from the radical signs which affect its unknown quantities. Examples 4, 5, 13, and 14 are thus found to furnish equations of the second degree, while examples 6 and 18 furnish equations of the fourth degree.

The above method of solving the equation $x^{2n}+px^n=q$ will not always give us *all* of the roots, and we must have recourse to different processes to discover the remaining roots. The subject will be resumed in Section XX.

PROBLEMS PRODUCING QUADRATIC EQUATIONS.

Prob. 1. It is required to find two numbers, such that their difference shall be 8, and their product 240.

Let $x =$ the least number.
Then will $x+8 =$ the greater.
And by the question $x(x+8)=x^2+8x=240$.
Therefore, $x=12$, the less number,
$x+8=20$, the greater.

Proof. $20-12=8$, the first condition.
$20\times12=240$, the second condition.

Prob. 2. The Receiving Reservoir at Yorkville is a rectangle, 60 rods longer than it is broad, and its area is 5500 square rods. Required its length and breadth?

Prob. 3. What two numbers are those whose difference is $2a$, and product b?

Ans. $a\pm\sqrt{a^2+b}$, and $-a\pm\sqrt{a^2+b}$.

Prob. 4. It is required to divide the number 60 into two such parts that their product shall be 864.

Let $x =$ one of the parts.
Then will $60-x =$ the other part.
And by the question, $x(60-x)=60x-x^2=864$.

The parts are 36 and 24. *Ans.*

Prob. 5. In a parcel which contains 52 coins of silver and copper, each silver coin is worth as many cents as there are copper coins, and each copper coin is worth as many cents as there are silver coins, and the whole are worth two dollars How many are there of each?

Prob. 6. What two numbers are those whose sum is $2a$, and product b?

Ans. $a+\sqrt{a^2-b}$, and $a-\sqrt{a^2-b}$.

Prob. 7. There is a number consisting of two digits whose sum is 10, and the sum of their squares is 58. Required the number.

Let $x=$ the first digit.

Then will $10-x=$ the second digit.

And $x^2+(10-x)^2=2x^2-20x+100=58$.

That is, $x^2-10x=-21$,

$x^2-10x+25=4$,

$x=5\pm2=7$, or 3.

Hence the number is 73, or 37.

The two values of x are the required digits whose sum is 10. It will be observed that we put x to represent the first digit, whereas we find it may equal the second as well as the first. The reason is, that we have here imposed a condition which does not enter into the equation. If x represent *either* of the required digits, then $10-x$ will represent the *other*, and hence the values of x found by solving the equation should give both digits. Beginners are very apt thus, in the statement of a problem, to impose conditions which do not appear in the equation.

The preceding example, and all others of the same class, may be solved without completing the square. Thus,

Let x represent the half difference of the two digits.

Then, according to the principle on page 67,

$5+x$ will represent the greater of the two digits,

$5-x$ " the less "

The square of $5+x$ is $25+10x+x^2$,

" $5-x$ $25-10x+x^2$,

The sum is $50+2x^2$, which, according to the problem, $=58$.

Hence $2x^2=8$,

or $x^2=4$,

and $x=\pm2$.

Therefore, $5+x=7$, the greater digit,

$5-x=3$, the less digit.

Prob. 8. Find two numbers such that the product of their sum and difference may be 5, and the product of the sum of their squares and the difference of their squares may be 65.

Prob. 9. Find two numbers such that the product of their sum and difference may be a, and the product of the sum of their squares and the difference of their squares may be ma.

Ans. $\sqrt{\frac{m+a}{2}}$; $\sqrt{\frac{m-a}{2}}$.

Prob. 10. A laborer dug two trenches, whose united length was 26 yards, for 356 shillings, and the digging of each of them cost as many shillings per yard as there were yards in its length. What was the length of each?

Ans. 10, or 16 yards.

Prob. 11. What two numbers are those whose sum is $2a$, and the sum of their squares is $2b$?

Ans. $a+\sqrt{b-a^2}$, and $a-\sqrt{b-a^2}$.

Prob. 12. A farmer bought a number of sheep for 80 dollars, and if he had bought four more for the same money, he would have paid one dollar less for each. How many did he buy?

Let x represent the number of sheep.

Then will $\frac{80}{x}$ be the price of each.

And $\frac{80}{x+4}$ would be the price of each, if he had bought four more for the same money.

But by the question we have

$$\frac{80}{x}=\frac{80}{x+4}+1.$$

Solving this equation, we obtain

$x=16$. *Ans.*

Prob. 13. A person bought a number of articles for a dollars. If he had bought $2b$ more for the same money, he would have paid c dollars less for each. How many did he buy?

Ans. $-b\pm\sqrt{\frac{2ab+b^2c}{c}}$.

Prob. 14. It is required to find three numbers such that the product of the first and second may be 15, the product of the first and third 21, and the sum of the squares of the second and third 74.

Ans. 3, 5, and 7.

Prob. 15. It is required to find three numbers such that the product of the first and second may be a, the product of the first and third b, and the sum of the squares of the second and third c.

$$Ans. \sqrt{\frac{a^2+b^2}{c}};\ a\sqrt{\frac{c}{a^2+b^2}};\ b\sqrt{\frac{c}{a^2+b^2}}.$$

Prob. 16. The sum of two numbers is 16, and the sum of their cubes 1072. What are those numbers?

Ans. 7 and 9.

Prob. 17. The sum of two numbers is $2a$, and the sum of their cubes is $2b$. What are the numbers?

$$Ans.\ a+\sqrt{\frac{b-a^3}{3a}} \text{ and } a-\sqrt{\frac{b-a^3}{3a}}.$$

Prob. 18. Two magnets, whose powers of attraction are as 4 to 9, are placed at a distance of 20 inches from each other. It is required to find, on the line which joins their centers, the point where a needle would be equally attracted by both, admitting that the intensity of magnetic attraction varies inversely as the square of the distance.

Ans. { 8 inches from the weakest magnet,
or -40 inches from the weakest magnet.

Prob. 19. Two magnets, whose powers are as m to n, are placed at a distance of a feet from each other. It is required to find, on the line which joins their centers, the point which is equally attracted by both.

Ans. { The distance from the magnet m is $\dfrac{a\sqrt{m}}{\sqrt{m}\pm\sqrt{n}}$.
The distance from the magnet n is $\dfrac{a\sqrt{n}}{\sqrt{m}\pm\sqrt{n}}$.

Prob. 20. A set out from C toward D, and traveled 6 miles an hour After he had gone 45 miles, B set out from D toward C, and went every hour $\frac{1}{20}$ of the entire distance; and after he had traveled as many hours as he went miles in one hour, he met A. Required the distance between the places C and D.

Ans. Either 100 miles, or 180 miles.

Prob. 21. A set out from C toward D, and traveled a miles per hour. After he had gone b miles, B set out from D toward C, and went every hour $\frac{1}{n}$th of the entire distance; and after he had traveled as many hours as he went miles in one hour, he met A. Required the distance between the places C and D.

$$Ans.\ n\left(\frac{n-a}{2}\pm\sqrt{\left(\frac{n-a}{2}\right)^2-b}\right).$$

Prob. 22. By selling my horse for 24 dollars, I lose as much per cent. as the horse cost me. What was the first cost of the horse?

Ans. 40, or 60 dollars.

QUADRATIC EQUATIONS CONTAINING TWO UNKNOWN QUANTITIES.

(186.) An equation containing two unknown quantities is said to be of the *second degree when it involves terms in which the sum of the exponents of the unknown quantities is equal to* 2, *but never exceeds* 2. Thus,

$$3x^2-4x+y^2=25,$$

and

$$7xy-4x+y=40,$$

are equations of the second degree.

The solution of two equations of the second degree containing two unknown quantities, generally involves the solution of an equation of the fourth degree containing one unknown quantity. Hence the principles hitherto established are not sufficient to enable us to solve *all* equations of this description. Yet there are particular cases in which they may be reduced either to pure or affected quadratics, and the roots determined in the ordinary manner.

(187.) When one of the equations is a simple equation, it is generally best to find an expression for the value of one of the unknown quantities from the simple equation, and substitute this value in the place of its equal in the other equation. The resulting equation will be of the second degree, and may be solved by the ordinary rules.

Ex. 1. Given $\left.\begin{aligned} x^2+3xy-\ y^2&=23 \\ x\ +2y&=\ 7 \end{aligned}\right\}$ to find the values of x and y.

From the second equation, we find

$$x=7-2y.$$

Whence $$x^2=49-28y+4y^2.$$

And, substituting this value in the first equation, we have

$$49-28y+4y^2+21y-6y^2-y^2=23,$$

a common quadratic equation, which may be solved in the usual manner.

Ans. $x=3$, and $y=2$.

Ex. 2. Given $\left.\begin{aligned} 2x^2+\ xy-5y^2&=20 \\ 2x\ -3y&=\ 1 \end{aligned}\right\}$ to find the values of x and y.

Ans. $x=5$, $y=3$.

Ex. 3. Given $\left.\begin{aligned} \frac{10x+y}{3}&=xy \\ 9y-9x&=18 \end{aligned}\right\}$ to find the values of x and y.

Ans. $x=2$, $y=4$.

(188.) When the same algebraic expression is involved to different powers, it is sometimes best to regard this expression as the unknown quantity.

Ex. 4. Given $\left.\begin{aligned} x^2+2xy+y^2\ +2x&=120-2y \\ xy-y^2&=8 \end{aligned}\right\}$ to find the values of x and y.

Here the first equation may be put under the form

$$(x+y)^2+2(x+y)=120,$$

where $x+y$ may be regarded as a single quantity, and, by completing the square, we shall find its value to be

either $$10, \text{ or } -12.$$

Proceeding now as in the last Article, we shall find

$$x=6, \text{ or } 9, \text{ or } -9\mp\sqrt{5},$$
$$y=4, \text{ or } 1, \text{ or } -3\pm\sqrt{5}.$$

Ex. 5. Given $\left.\begin{aligned} 4xy&=96-x^2y^2 \\ x+y&=6 \end{aligned}\right\}$ to find the values of x and y.

Here we may regard xy as the unknown quantity, and we shall find its value from the first equation to be

either $$8, \text{ or } -12.$$

Proceeding again as in the former Article, we shall find

$$x=2, \text{ or } 4, \text{ or } 3\pm\sqrt{21},$$
$$y=4, \text{ or } 2, \text{ or } 3\mp\sqrt{21}.$$

Ex. 6. Given $\left.\begin{array}{l}\dfrac{x^2}{y^2}+\dfrac{4x}{y}=\dfrac{85}{9}\\ x-y=2\end{array}\right\}$ to find the values of x and y.

Here $\dfrac{x}{y}$ may be treated as the unknown quantity, and we shall find its value to be

either $$\frac{5}{3}, \text{ or } -\frac{17}{3}.$$

From which we find

$$x=5, \text{ or } \frac{17}{10},$$
$$y=3, \text{ or } -\frac{3}{10}.$$

(189.) When the sum of the dimensions of the unknown quantities is the same in every term of the two equations, it is sometimes best to substitute for one of the unknown quantities the product of the other by a third unknown quantity.

Ex. 7. Given $\left.\begin{array}{l}x^2+xy=56\\ xy+2y^2=60\end{array}\right\}$ to find the values of x and y.

Here, if we assume $x=vy$, we shall have

$$v^2y^2+vy^2=56,$$
$$v\,y^2+2y^2=60.$$

From the first of these equations,

$$y^2=\frac{56}{v^2+v},$$

and from the second, $$y^2=\frac{60}{v+2};$$

therefore, $$\frac{60}{v+2}=\frac{56}{v^2+v}.$$

From which, after completing the square, we obtain

$$v=\frac{4}{3}, \text{ or } -\frac{7}{5}.$$

Substituting either of these values in one of the preceding expressions for y^2, we shall obtain the values of y; and since $x=vy$, we may easily obtain the values of x.

Ans. $\left\{\begin{array}{l}x=\pm14, \text{ or } \pm4\sqrt{2},\\ y=\mp10, \text{ or } \pm3\sqrt{2}.\end{array}\right.$

Ex. 8. Given $\left.\begin{array}{r}2x^2+3xy+\ y^2=20\\ 5x^2\ +4y^2=41\end{array}\right\}$ to find the values of x and y.

If we assume $x=vy$, we shall find

$$v=\frac{1}{3}, \text{ or } \frac{13}{2},$$

whence, as before, we shall obtain

$$Ans. \begin{cases} x=\pm 1, \text{ or } \dfrac{\pm 13}{\sqrt{21}}, \\ y=\pm 3, \text{ or } \dfrac{\pm 2}{\sqrt{21}}. \end{cases}$$

Ex. 9. Given $\left.\begin{array}{r}x^2+xy=77\\ xy-y^2=12\end{array}\right\}$ to find the values of x and y

If we assume $x=vy$, we shall find

$$v=\frac{7}{4}, \text{ or } \frac{11}{3},$$

whence, as before,

$$Ans. \begin{cases} x=\pm 7, \text{ or } \dfrac{\pm 11}{\sqrt{2}}, \\ y=\pm 4, \text{ or } \dfrac{\pm 3}{\sqrt{2}}. \end{cases}$$

(190.) When the unknown quantities in each equation are similarly involved, it is sometimes best to substitute for the unknown quantities the sum and difference of two other quantities, or the sum and product of two other quantities.

Ex. 10. Given $\left.\begin{array}{r}\dfrac{x^2}{y}+\dfrac{y^2}{x}=18\\ x+y=12\end{array}\right\}$ to find the values of x and y.

Here let us assume

$$x=z+v,$$
$$y=z-v.$$

Then, by adding these two equations together, we shall have

$$x+y=2z=12, \text{ or } z=6;$$

that is, $x=6+v$, and $y=6-v$.

But, from the first equation, we find

$$x^3+y^3=18xy.$$

Substituting the preceding values of x and y in this equation, and reducing, we obtain

$$432+36v^2=648-18v^2.$$

Whence $v=\pm 2.$

Therefore, $x=4$, or 8,

and $y=8$, or 4.

Ex. 11. Given $\left.\begin{array}{r} x^5+y^5=3368 \\ x+y=\quad 8 \end{array}\right\}$ to find the values of x and y.

Ans. $\left\{\begin{array}{l} x=3, \text{ or } 5, \\ y=5, \text{ or } 3. \end{array}\right.$

Ex. 12. Given $\left.\begin{array}{r} x^3+y^3=341 \\ x^2y+xy^2=330 \end{array}\right\}$ to find the values of x and y.

Ans. $\left\{\begin{array}{l} x=5, \text{ or } 6, \\ y=6, \text{ or } 5. \end{array}\right.$

PROBLEMS.

1. Divide the number 100 into two such parts, that the sum of their square roots may be 14.

Ans. 64 and 36.

2. Divide the number a into two such parts, that the sum of their square roots may be b.

Ans. $\frac{a}{2}\pm\frac{b}{2}\sqrt{2a-b^2}.$

3. The sum of two numbers is 8, and the sum of their fourth powers is 706. What are the numbers?

Ans. 3 and 5.

4. The sum of two numbers is $2a$, and the sum of their fourth powers is $2b$. What are the numbers?

Ans. $a\pm\sqrt{-3a^2+\sqrt{8a^4+b}}.$

5. The sum of two numbers is 6, and the sum of their fifth powers is 1056. What are the numbers?

Ans. 2 and 4.

6. The sum of two numbers is $2a$, and the sum of their fifth powers is b. What are the numbers?

Ans. $a\pm\sqrt{\sqrt{\frac{b}{10a}+\frac{4a^4}{5}}-a^2}.$

7. What two numbers are those whose product is 120; and if the greater be increased by 8 and the less by 5, the product of the two numbers thus obtained shall be 300?

Ans. 12 and 10, or 16 and 7.5.

8. What two numbers are those whose product is a; and if the greater be increased by b and the less by c, the product of the two numbers thus obtained shall be d?

$$Ans.\ \frac{m}{2}\pm\sqrt{\frac{m^2}{4}-\frac{ab}{c}},\ \text{and}\ \frac{a}{\frac{m}{2}\pm\sqrt{\frac{m^2}{4}-\frac{ab}{c}}},$$

where $m=\dfrac{d-a-bc}{c}$.

9. Find two numbers such that their sum, their product, and the difference of their squares may be all equal to one another.

$$Ans.\ \frac{3}{2}+\sqrt{\frac{5}{4}},\ \text{and}\ \frac{1}{2}+\sqrt{\frac{5}{4}},$$

that is, 2.618, and 1.618, nearly.

10. Divide the number 100 into two such parts, that their product may be equal to the difference of their squares.

Ans. 38.197, and 61.803.

11. Divide the number a into two such parts, that their product may be equal to the difference of their squares.

$$Ans.\ \frac{3a\pm a\sqrt{5}}{2}\ \text{and}\ \frac{-a\mp a\sqrt{5}}{2}$$

DISCUSSION OF THE GENERAL EQUATION OF THE SECOND DEGREE.

(191.) We have seen, *Art.* 181, that every equation of the second degree may be reduced to the form

$$x^2+px=q,$$

where p and q represent known quantities, either positive or negative, integral or fractional.

The value of x in this equation is

either $$x=-\frac{p}{2}+\sqrt{q+\frac{p^2}{4}},$$

or $$x=-\frac{p}{2}-\sqrt{q+\frac{p^2}{4}}.$$

And, since these values necessarily result from the general equation, we infer,

PROPERTY I.

Every equation of the second degree has two roots, and only two.

A *root* of an equation is such a number as, being substituted for the unknown quantity, will satisfy the equation.

This principle has been often exemplified in the preceding pages. Two values have uniformly been found for x, although both values may not be applicable to the problem which furnishes the equation. This property will be found demonstrated in a general manner in *Art.* 294.

(192.) If we multiply

$$x+\frac{p}{2}-\sqrt{q+\frac{p^2}{4}}=0,$$

by
$$x+\frac{p}{2}+\sqrt{q+\frac{p^2}{4}}=0,$$

we shall obtain $\qquad x^2+px-q=0,$

which was the equation originally proposed.

Hence,

PROPERTY II.

Every equation of the second degree, whose roots are a *and* b, *may be resolved into the two factors* x—a *and* x—b.

Ex. 1. Thus the equation

$$x^2-10x+16=0,$$

may be resolved into the factors $\begin{cases} x-8=0, \\ x-2=0, \end{cases}$

where 8 and 2 are the roots of the given equation.

It is also obvious that if a is a root of an equation of the second degree, this equation must be divisible by $x-a$. Thus the preceding equation is divisible by $x-8$, giving the quotient $x-2$.

Ex. 2. The roots of the equation

$$x^2+6x+8=0,$$

are -2 and -4. Resolve it into its factors.

Ex. 3. The roots of the equation

$$x^2+6x-27=0,$$

are $+3$ and -9. Resolve it into its factors.

Ex. 4. The roots of the equation

$$x^2-2x-24=0,$$

are $+6$ and -4. Resolve it into its factors.

(193.) If we add together the two values of x in the general equation of the second degree, the radical parts having opposite signs disappear, and we obtain

$$-\frac{p}{2}-\frac{p}{2}=-p.$$

Hence,

PROPERTY III.

The algebraic sum of the two roots is equal to the coefficient of the second term of the equation, taken with a contrary sign.

Thus, in *Ex.* 1, page 145,

$$x^2-10x=-16,$$

the two roots are 8 and 2, whose sum is $+10$, the coefficient of x taken with a contrary sign.

In the equation

$$x^2+6x=-8,$$

the two roots are -2 and -4.

In the equation

$$x^2+16x=-60,$$

the two roots are -6 and -10.

If the two roots are equal numerically, but have opposite signs, their sum is zero, and the second term of the equation vanishes. Thus the two roots of the equation $x^2=16$, are $+4$ and -4, whose sum is zero. This equation may be written

$$x^2+0x=16.$$

(194.) If we multiply together the two values of x (observing that the product of the sum and difference of two quantities is equal to the difference of their squares), we obtain

$$\frac{p^2}{4}-\left(q+\frac{p^2}{4}\right)=-q.$$

Hence,

PROPERTY IV.

The product of the two roots is equal to the second member of the equation, taken with a contrary sign.

Thus, in the equation

$$x^2-10x=-16,$$

the product of the two roots 8 and 2 is +16, which is equal to the second member of the equation taken with a contrary sign.

So, also, in the equation

$$x^2+6x=27,$$

whose two roots are +3 and −9, their product is −27.

The two last properties enable us readily to form an equation when its roots are known.

Ex. 1. Let it be required to form the equation whose roots are 2 and 8.

According to Property III., the coefficient of the second term of the equation must be −10; and, from Property IV. the second member of the equation must be −16. Hence the equation is

$$x^2-10x=-16.$$

Ex. 2. Form the equation whose roots are 3 and 5.

Ex. 3. Form the equation whose roots are −4 and −7.

Ex. 4. Form the equation whose roots are 5 and −9.

Ex. 5. Form the equation whose roots are −6 and +11.

REAL AND IMAGINARY VALUES OF THE UNKNOWN QUANTITY.

(195.) The values of x in the general equation of the second degree are

$$x=-\frac{p}{2}\pm\sqrt{q+\frac{p^2}{4}}.$$

Values of the unknown quantity which are *not imaginary* are, for the sake of distinction, called *real.*

Since $\frac{p^2}{4}$, being a square, is positive for all real values of p, it follows that the expression $q+\frac{p^2}{4}$ can only be rendered negative by the sign of q.

When q is positive, or when q is negative and numerically less than $\frac{p^2}{4}$, then will $q+\frac{p^2}{4}$ be positive, and, consequently, $\sqrt{q+\frac{p^2}{4}}$ will be *real.* This happens in nearly all the preceding examples.

When q is negative, and numerically greater than $\frac{p^2}{4}$, then $q+\frac{p^2}{4}$ will be negative, and, consequently, $\sqrt{q+\frac{p^2}{4}}$ will be *imaginary.* This happens in *Ex.* 5, page 146.

CASE I.

When $\sqrt{q+\frac{p^2}{4}}$ *is real.*

1. When, in the equation $x^2+px=q$, p is negative, and $\frac{p}{2}$ is numerically greater than $\sqrt{q+\frac{p^2}{4}}$, *both values of* x *will be real and positive.*

This happens in the equation

$$x^2-6x=-8,$$

whose two roots are 4 and 2.

Also in the equation

$$x^2-10x=-16,$$

whose two roots are 8 and 2.

2. When p is positive, and $\frac{p}{2}$ is numerically greater than $\sqrt{q+\frac{p^2}{4}}$, *both values of* x *will be real and negative.*

This happens in the equation

$$x^2+6x=-8,$$

whose two roots are -2 and -4.

Also in the equation

$$x^2+16x=-60,$$

whose two roots are -6 and -10.

3. When $\frac{p}{2}$ is numerically less than $\sqrt{q+\frac{p^2}{4}}$, *both values of* x *will be real, the one positive and the other negative.*

This happens in the equation

$$x^2+6x=27,$$

whose roots are $+3$ and -9.

Also in the equation

$$x^2-2x=24,$$

whose roots are $+6$ and -4.

CASE II.

(196.) *When* $\sqrt{q+\frac{p^2}{4}}$ *is imaginary.*

In this case, *both values of* x *are imaginary.*

This happens in the equation

$$x^2-8x=-18,$$

whose roots are $4\pm\sqrt{-2}$.

We will now prove that in this case the conditions of the question are *incompatible* with each other, and therefore the values of x *ought* to be imaginary. The demonstration depends upon the following principle:

The greatest product which can be obtained by dividing a number into two parts and multiplying them together, is the square of half that number.

Let p = the given number,
and d = the difference of the parts.

Then, from page 67, $\frac{p}{2}+\frac{d}{2}$ = the greater part,

$\frac{p}{2}-\frac{d}{2}$ = the less part,

and $$\frac{p^2}{4}-\frac{d^2}{4} = \text{the product of the parts.}$$

Now, since p is a given quantity, it is plain that this expression will be the greatest possible when $d=0$; that is, $\frac{p^2}{4}$ is the greatest product, which is the square of $\frac{p}{2}$, half the given number.

For example, let 12 be the number to be divided.

We have $12=1+11$; and $11\times1=11$.
$12=2+10$; and $10\times2=20$.
$12=3+9$; and $9\times3=27$.
$12=4+8$; and $8\times4=32$.
$12=5+7$; and $7\times5=35$.
$12=6+6$; and $6\times6=36$.

We here see that the smaller the difference of the two parts, the greater is their product; and this product is greatest when the two parts are equal.

Now, in the equation

$$x^2-px=-q,$$

p is the sum of the two roots, and q is their product. Therefore q can never be greater than $\frac{p^2}{4}$.

If, then, any problem furnishes an equation in which q is negative, and greater than $\frac{p^2}{4}$, we infer that the conditions of the question are incompatible with each other.

Thus, in the example

$$x^2-6x=-10,$$

$\frac{p^2}{4}=9$, which is numerically less than q. The equation requires us to divide the number 6 into two parts whose product shall be 10, which is an impossibility; and, accordingly, in solving the equation, we obtain *imaginary* values for x.

Hence *an imaginary root indicates an absurdity in the proposed question which furnished the equation.*

Suppose it is required to divide 8 into two such parts that their product shall be 18.

Let $x =$ one of the parts,
and $8-x =$ the other.

Then, by the conditions,

$$x(8-x)=18.$$

Whence $x^2-8x=-18.$

This equation, solved by the usual method, gives

$$x=4\pm\sqrt{-2}, \text{ an imaginary expression.}$$

Hence we infer that it is *impossible* to find two numbers whose sum is 8, and product 18. This is obvious from the Proposition above demonstrated, from which it appears that 16 is the greatest product which can be obtained by dividing 8 into two parts, and multiplying them together.

(197.) When q is negative, and numerically equal to $\frac{p^2}{4}$, the radical part of both values of x becomes zero, and both values of x reduce to $-\frac{p}{2}$. *The two roots are then said to be equal.*

Thus, in the equation

$$x^2-6x=-9,$$

the two roots are 3 and 3.

We say that in this case the equation has two roots, because it is the product of the two factors, $x-3=0$, and $x-3=0$.

DISCUSSION OF PARTICULAR PROBLEMS.

(198.) In discussing particular problems which involve equations of the second degree, we meet with all the different cases which are presented by equations of the first degree, and some peculiarities besides. We may therefore have,

1. Positive values of x.
2. Negative values.
3. Values of the form of $\frac{0}{A}$.
4. Values of the form of $\frac{A}{0}$.
5. Values of the form of $\frac{0}{0}$.

All these different cases are presented by Problem 19

page 155, when we make different suppositions upon the values of a, m, and n; but we need not dwell upon them here.

The peculiarities exhibited by equations of the second degree are,

6. Double values of x.

7. Imaginary values.

We will consider the last two cases.

(199.) *Double values of the unknown quantity.*

We have seen that every equation of the second degree has two roots. Sometimes both of these values are applicable to the problem which furnishes the equation. Thus, in Problem 20, page 155, we obtain either 100 or 180 miles for the distance between the places C and D.

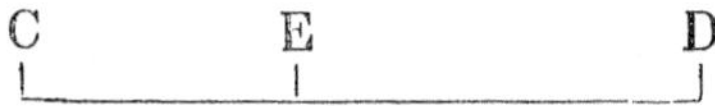

Let E represent the situation of A when B sets out on his journey. Then, if we suppose CD equals 100 miles, ED will equal 55 miles, of which A will travel 30 miles (being 6 miles an hour for 5 hours), and B will travel 25 miles (being 5 miles an hour for 5 hours).

If we suppose CD equals 180 miles, ED will equal 135 miles, of which A will travel 54 miles (being 6 miles an hour for 9 hours), and B will travel 81 miles (being 9 miles an hour for 9 hours).

This problem, therefore, admits of two positive answers, both equally applicable to the question.

Problem 22, page 156, is of the same kind; and another will be found on page 193.

In Problem 18, page 155, one of the values of x is positive, and the other negative.

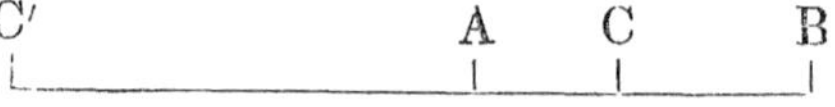

Let the weakest magnet be placed at A, and the strongest at B; then C will represent the situation of a needle equally attracted by both magnets. According to the first value, the distance AC=8 inches, and CB=12. Now at the distance of 8 inches, the attraction of the weakest magnet will be represented by $\frac{4}{8^2}$; and at the distance of 12 inches, the attraction of the other magnet will be represented by $\frac{9}{12^2}$, and these two powers are equal; for

$$\frac{4}{8^2}=\frac{9}{12^2}.$$

But there is another point, C′, which equally satisfies the conditions of the question; and this point is 40 inches to the left of A, and therefore 60 inches to the left of B; for

$$\frac{4}{40^2}=\frac{9}{60^2}.$$

(200.) *Imaginary values of the unknown quantity.*

We have seen that an imaginary root indicates an absurdity in the proposed question which furnished the equation.

In several of the preceding problems, the values of x become imaginary in particular cases.

When will the values of x in Problem 6, page 153, be imaginary?

Ans. When $b>a^2$

What is the absurdity involved in this supposition?

Ans. It is absurd to suppose that the product of two numbers can be greater than the square of half their sum.

When will the values of x in Problem 11, page 154, be imaginary?

Ans. When $a^2>b$; or $(2a)^2>4b$.

What is the absurdity of this supposition?

Ans. The square of the sum of two numbers can not be greater than twice the sum of their squares.

When will the values of x in Problem 17, page 155, be imaginary?

Ans. When $a^3>b$; or $(2a)^3>8b$

What is the absurdity of this supposition?

Ans. The cube of the sum of two numbers can not be greater than four times the sum of their cubes.

When will the values of x in Problem 4, page 140, be imaginary, and what is the absurdity of this supposition?

SECTION XIII.

RATIO AND PROPORTION.

(201.) Numbers may be compared in two ways: either by means of their *difference*, or by their *quotient*. We may inquire *how much* one quantity is greater than another; or, *how many times* the one contains the other. One is called Arithmetical, and the other Geometrical Ratio.

The difference between two numbers is called their *Arithmetical Ratio*. Thus, the arithmetical ratio of 9 to 7 is $9-7$, or 2; and if a and b designate two numbers, their arithmetical ratio is represented by $a-b$.

Numbers are more generally compared by means of quotients; that is, by inquiring how many times one number contains another. The quotient of one number divided by another is called their *Geometrical Ratio*. The term Ratio, when used without any qualification, is always understood to signify a geometrical ratio, and we shall confine our attention to ratios of this description.

(202.) By the *ratio* of two numbers, then, we mean *the quotient which arises from dividing one of these numbers by the other*.

Thus, the ratio of 12 to 4 is represented by $\frac{12}{4}$, or 3.

The ratio of 5 to 2 is $\frac{5}{2}$, or 2.5.

The ratio of 1 to 3 is $\frac{1}{3}$, or .333, &c.

We here perceive that the value of a ratio can not always be expressed exactly in decimals; but, by taking a sufficient number of terms, we can approach as nearly as we please to the true value.

If a and b designate two numbers, the ratio of a to b is the quotient arising from dividing a by b, and may be represented by writing them $a:b$, or $\frac{a}{b}$. The first term, a, is called the *antecedent* of the ratio; the last term, b, is called the *consequent* of the ratio.

Hence it appears that the theory of ratios is included in the theory of fractions, and a ratio may be considered as *a fraction whose numerator is the antecedent, and whose denominator is the consequent.*

(203.) When the antecedent of a ratio is greater than the consequent, the ratio is called *a ratio of greater inequality;* as, $\frac{5}{3}$, $\frac{12}{4}$. When the antecedent is less than the consequent, it is called *a ratio of less inequality;* as, $\frac{2}{3}$, $\frac{5}{9}$. When the antecedent and consequent are equal, it is called *a ratio of equality;* as, $\frac{3}{3}$, $\frac{8}{8}$. It is plain that a ratio of equality may always be represented by unity.

(204.) When the corresponding terms of two or more simple ratios are multiplied together, the ratios are said to be *compounded.* Thus, the ratio of $\frac{a}{b}$, compounded with the ratio of $\frac{c}{d}$, becomes $\frac{ac}{bd}$.

When a ratio is compounded with itself, the result is called *a duplicate ratio.* Thus, the duplicate ratio of $\frac{2}{3}$ is $\frac{4}{9}$; and the duplicate ratio of $\frac{a}{b}$ is $\frac{a^2}{b^2}$.

A ratio compounded of three equal ratios is called *a triplicate ratio.* Thus, the triplicate ratio of $\frac{2}{3}$ is $\frac{8}{27}$; and the triplicate ratio of $\frac{a}{b}$ is $\frac{a^3}{b^3}$.

The ratio of the *square roots* of two quantities is called a *subduplicate ratio.* Thus, the subduplicate ratio of $\frac{4}{9}$ is $\frac{2}{3}$; and the subduplicate ratio of $\frac{a}{b}$ is $\frac{\sqrt{a}}{\sqrt{b}}$.

The ratio of the *cube roots* of two quantities is called a *subtriplicate ratio.* Thus, the subtriplicate ratio of $\frac{8}{27}$ is $\frac{2}{3}$; and the subtriplicate ratio of $\frac{a}{b}$ is $\frac{\sqrt[3]{a}}{\sqrt[3]{b}}$.

(205.) *If the terms of a ratio are both multiplied, or both divided by the same quantity, the value of the ratio remains unchanged.*

The ratio of a to b is represented by the fraction $\frac{a}{b}$, and the value of a fraction is not changed if we multiply or divide both numerator and denominator by the same quantity. Thus,

$$\frac{a}{b}=\frac{ma}{mb}=\frac{\frac{a}{n}}{\frac{b}{n}};$$

or

$$a:b=ma:mb=\frac{a}{n}:\frac{b}{n}.$$

(206.) Ratios are compared with each other by reducing the fractions which represent them to a common denominator.

In order to ascertain whether the ratio of 2 to 7 is greater or less than that of 3 to 8, we represent these ratios by the fractions $\frac{2}{7}$ and $\frac{3}{8}$, and reduce them to a common denominator. They thus become

$$\frac{16}{56} \text{ and } \frac{21}{56};$$

and, since the latter of these is the greatest, we infer that the ratio of 2 to 7 is less than the ratio of 3 to 8.

(207.) A ratio of *greater inequality is diminished,* and a ratio of *less inequality is increased,* by adding the same quantity to both terms.

Thus,
$$\frac{3}{2}>\frac{3+1}{2+1}, \text{ or } \frac{4}{3};$$

$$\frac{2}{3}<\frac{2+1}{3+1}, \text{ or } \frac{3}{4}.$$

To prove the proposition generally, let $\frac{a}{b}$ represent any ratio, and let x be added to each of its terms. The two ratios will then be

$$\frac{a}{b} \text{ and } \frac{a+x}{b+x},$$

which, reduced to a common denominator, become

$$\frac{ab+ax}{b(b+x)}, \frac{ab+bx}{b(b+x)}.$$

Now if $a>b$, that is, if $\frac{a}{b}$ is a ratio of greater inequality, then, since ax is greater than bx, the first of these fractions is greater than the second, and therefore $\frac{a}{b}$ is *diminished* by the addition of the same quantity to each of its terms.

But if $a<b$, that is, if $\frac{a}{b}$ is a ratio of less inequality, then, since ax is less than bx, the first of the above fractions is less than the second, and therefore $\frac{a}{b}$ is *increased* by the addition of the same quantity to each of its terms.

(208.) If, in a series of ratios, the consequent of each is the antecedent of the following ratio, then *the ratio of the first antecedent to the last consequent is equal to that which is compounded of all the intervening ratios.*

Let the proposed ratios be

$$\frac{a}{b}, \frac{b}{c}, \frac{c}{d}, \frac{d}{e}, \frac{e}{f}.$$

Compounding them by *Art.* 204, we obtain

$$\frac{abcde}{bcdef},$$

which, being divided by $bcde$, reduces to

$$\frac{a}{f}.$$

PROPORTION.

(209.) *Proportion is an equality of ratios.*

Thus, if a, b, c, d are four quantities, such that a, when divided by b, gives the same quotient as c when divided by d, then a, b, c, d are called *proportionals*, and we say that a is to b as c is to d; and this is expressed by writing them thus:

$$a : b :: c : d,$$

or

$$a : b = c : d,$$

or

$$\frac{a}{b} = \frac{c}{d}.$$

So, also, 3, 4, 9, 12 are proportionals; that is,

$$3 : 4 :: 9 : 12$$

or

$$\frac{3}{4} = \frac{9}{12}.$$

In ordinary language, the terms *ratio* and *proportion* are confounded with each other. Thus, two quantities are said to be in the proportion of 3 to 5, instead of the ratio of 3 to 5. A ratio subsists between *two* quantities, a proportion only between *four*. Ratio is the quotient arising from dividing one quantity by another; *two equal ratios form a proportion.*

(210.) In the proportion

$$a : b :: c : d,$$

a, b, c, d are called the *terms* of the proportion. The first and last terms are called the *extremes*, the second and third the *means*. The first term is called the *first antecedent*, the second term the *first consequent*, the third term the *second antecedent*, and the fourth term the *second consequent.*

The word *term*, when applied to a proportion, is used in a slightly different sense from that explained in *Art.* 27. The terms of a proportion may be polynomials. Thus,

$$a+b : c+d :: e+f : g+h.$$

(211.) When the second and third terms of a proportion are identical, this quantity is called a *mean proportional* between the other two. Thus, if we have three quantities, a, b, c, such that

$$a : b :: b : c,$$

then b is called a *mean proportional* between a and c, and c is called a *third proportional* to a and b.

If, in a series of proportional magnitudes, each consequent is identical with the next antecedent, these quantities are said to be in *continued proportion.* Thus, if we have a, b, c, d, e, f such that

$$a : b :: b : c :: c : d :: d : e :: e : f,$$

or

$$\frac{a}{b}=\frac{b}{c}=\frac{c}{d}=\frac{d}{e}=\frac{e}{f},$$

the quantities a, b, c, d, e, f are in continued proportion.

(212.) If four quantities are proportional, *the product of the extremes is equal to the product of the means.*

Let $$a : b :: c : d.$$

Then will $$ad=bc.$$

For, since the four quantities are proportional,

$$\frac{a}{b}=\frac{c}{d}.$$

Multiplying each of these equals by bd, the expression becomes

$$\frac{abd}{b}=\frac{bcd}{d},$$

or $$ad=bc.$$

Thus, if $$3 : 4 :: 9 : 12,$$

then $$3\times12=4\times9.$$

(213.) Conversely, if the product of two quantities is equal to the product of two others, *the first two quantities may be made the extremes, and the other two the means of a proportion.*

Let $$ad=bc.$$

Then will $$a : b :: c : d.$$

For, since $$ad=bc,$$

dividing each of these equals by bd, the expression becomes

$$\frac{a}{b}=\frac{c}{d}, \text{ or } \frac{c}{d}=\frac{a}{b};$$

that is, $$a : b :: c : d, \text{ or } c : d :: a : b.$$

Thus, if $$3\times12=4\times9,$$

then $$3 : 4 :: 9 : 12,$$

or $$9 : 12 :: 3 : 4.$$

(214.) The preceding proposition is called the *test* of proportions, and any change may be made in the form of a proportion which is consistent with the application of this test. In order, then, to decide whether four quantities are proportional, we must compare the product of the extremes with the product of the means.

Thus, to determine whether 5, 6, 7, 8 are proportional, we multiply 5 by 8, and obtain 40. Multiplying 6 by 7, we obtain 42. As these two products are *not* equal, we conclude that the numbers 5, 6, 7, 8 are *not proportional.*

Again, take the numbers 5, 6, 10, 12. The product of 5 by 12 is 60, and the product of 6 by 10 is also 60. Hence these numbers are proportional; that is,

$$5 : 6 :: 10 : 12.$$

(215.) If three quantities are in continued proportion, *the product of the extremes is equal to the square of the mean.*

If $a : b :: b : c$.

Then, by *Art.* 212, $ac = bb$, which is equal to b^2.

Conversely, if the product of two quantities is equal to the square of a third, *the last quantity is a mean proportional between the other two.*

Thus, let $ac = b^2$.

Dividing these equals by bc, we obtain

$$\frac{a}{b} = \frac{b}{c},$$

or $a : b :: b : c$.

Thus, if $4 : 6 :: 6 : 9$,

then $4 \times 9 = 6^2$.

And conversely, if $4 \times 9 = 6^2$,

then 6 is a mean proportional between 4 and 9.

EXAMPLES.

1. Given the first three terms of a proportion, 24, 15, and 40, to find the fourth term.

2. Given the first three terms of a proportion, $3ab^3$, $4a^2b^2$, and $9a^3b$, to find the fourth term.

3. Given the last three terms of a proportion, $4a^3b^5$, $3a^3b^3$, and $2a^5b$, to find the first term.

4. Given the first, second, and fourth terms of a proportion, $5y^4$, $7x^2y^3$, and $21x^6y$, to find the third term.

5. Given the first, third, and fourth terms of a proportion, $a+b$, a^2-b^2, and $(a-b)^2$, to find the second term.

(216.) *Ratios that are equal to the same ratio are equal to each other.*

Let $a:b::x:y,$ and $c:d::x:y,$ then will $a:b::c:d.$

For, since $a:b::x:y,$

we have $$\frac{a}{b}=\frac{x}{y}.$$

And since $c:d::x:y,$

we have $$\frac{c}{d}=\frac{x}{y}.$$

Therefore, $$\frac{a}{b}=\frac{c}{d},$$

and hence $a:b::c:d.$

(217.) If four quantities are proportional, they will be proportional by *alternation;* that is, *the first will have the same ratio to the third that the second has to the fourth.*

Let $a:b::c:d,$

then will $a:c::b:d.$

For since $a:b::c:d,$

by *Art.* 212, $ad=bc,$

and since $ad=bc,$

by *Art.* 213, $a:c::b:d.$

(218.) If four quantities are proportional, they will be proportional by *inversion;* that is, *the second will have to the first the same ratio that the fourth has to the third.*

Let $a:b::c:d;$

then will $b.a::d:c.$

For since $a:b::c:d,$

by *Art.* 212, $ad=bc,$

or $bc=ad.$

Therefore, by *Art.* 213, $b:a::d:c.$

(219.) If four quantities are proportional, they will be pro portional by *composition;* that is, *the sum of the first and second will have to the second the same ratio that the sum of the third and fourth has to the fourth.*

Let $a:b::c:d;$

then will $a+b:b::c+d:d.$

For since $a:b::c:d,$

we have $\frac{a}{b}=\frac{c}{d}.$

Add unity to each of these equals, and we have

$$\frac{a}{b}+1=\frac{c}{d}+1, \text{ or } \frac{a+b}{b}=\frac{c+d}{d};$$

that is, $a+b:b::c+d:d.$

(220.) If four quantities are proportional, they will be proportional by *division;* that is, *the difference of the first and second will have to the second the same ratio that the difference of the third and fourth has to the fourth.*

Let $a:b::c:d;$

then will $a-b:b::c-d:d.$

For since $a:b::c:d,$

we have $\frac{a}{b}=\frac{c}{d}.$

Subtract unity from each of these equals, and we have

$$\frac{a}{b}-1=\frac{c}{d}-1, \text{ or } \frac{a-b}{b}=\frac{c-d}{d};$$

that is, $a-b:b::c-d:d.$

(221.) If four quantities are proportional, they will be proportional by *conversion;* that is, *the first will have to the difference of the first and second the same ratio that the third has to the difference of the third and fourth.*

Let $a:b::c:d;$

then will $a:a-b::c:c-d.$

For since $a:b::c:d,$

by inversion, $b:a::d:c;$

whence $\frac{b}{a}=\frac{d}{c}.$

Subtract each of these equals from unity, and we have

$$1-\frac{b}{a}=1-\frac{d}{c}, \text{ or } \frac{a-b}{a}=\frac{c-d}{c};$$

that is, $a-b : a :: c-d : c,$

or inversely, $a : a-b :: c : c-d.$

(222.) If four quantities are proportional, *the sum of the first and second will have to their difference the same ratio that the sum of the third and fourth has to their difference.*

Let $a : b :: c : d;$

then will $a+b : a-b :: c+d : c-d.$

For since $a : b :: c : d,$

by composition, $a+b : b :: c+d : d,$

and by alternation, $a+b : c+d :: b : d.$

Also, since $a : b :: c : d,$

by division, $a-b : b :: c-d : d,$

and by alternation, $a-b : c-d :: b : d.$

Hence, by equality of ratios,

$$a+b : a-b : c+d : c-d.$$

(223.) If four quantities are proportional, *like powers or roots of these quantities will also be proportional.*

Let $a : b :: c : d;$

then will $a^n : b^n :: c^n : d^n.$

For since $a : b :: c : d,$

we have $\frac{a}{b}=\frac{c}{d}.$

Raising each of these equals to the nth power, we obtain

$$\frac{a^n}{b^n}=\frac{c^n}{d^n};$$

that is, $a^n : b^n :: c^n : d^n,$

where n may be either a whole number or a fraction.

(224.) If there is any number of proportional quantities all having the same ratio, *the first will have to the second the same ratio that the sum of all the antecedents has to the sum of all the consequents.*

Let a, b, c, d, e, f be any number of proportional quantities such that

$$a : b :: c : d :: e : f,$$

then will $a : b :: a+c+e : b+d+f.$

For since $a : b :: c : d$,
we have $ad=bc$;
and since $a : b :: e : f$,
we have $af=be$.

To these equals add $ab=ba$,
and we obtain $a(b+d+f)=b(a+c+e)$.

Hence, by *Art.* 213, $a : b :: a+c+e : b+d+f$.

(225.) If three quantities are in continued proportion, *the first will have to the third the duplicate ratio of that which it has to the second.*

Let $a : b :: b : c$.
Then $a : c :: a^2 : b^2$.

For since $a : b :: b : c$,
by *Art.* 212, $ac=b^2$.

Multiplying each of these equals by a, we obtain

$$a^2c=ab^2;$$

that is, $a^2\times c=a\times b^2$.

Resolving this equation into a proportion by *Art.* 213, we have

$$a : c :: a^2 : b^2.$$

(226.) If four quantities are in continued proportion, *the first will have to the fourth the triplicate ratio of that which it has to the second.*

Let a, b, c, d be four quantities in continued proportion, so that

$$a : b :: b : c :: c : d,$$

then will $a : d :: a^3 : b^3$.

For since $a : b :: c : d$,
we have $ad=bc$;
and since $a : b :: b : c$,
we have $ac=b^2$.

Multiplying these equals by ab, we obtain

$$a^3(bdc)=b^3(abc),$$

or $a^3\times d=b^3\times a$.

Hence, by *Art.* 213, $a : d :: a^3 : b^3$.

(227.) If there are two sets of proportional quantities, *the products of the corresponding terms will be proportional.*

Let $a : b :: c : d,$

and $e : f :: g : h.$

Then will $ae : bf :: cg : dh.$

For, since $a : b :: c : d,$

by *Art.* 212, $ad = bc.$

And since $e : f :: g : h,$

by *Art.* 212, $eh = fg.$

Multiplying these equals together, we have

$$ae \times dh = bf \times cg.$$

Hence, by *Art.* 213, $ae : bf :: cg : dh.$

(228.) Three quantities are said to be in *harmonical* proportion when *the first is to the third as the difference between the first and second is to the difference between the second and third.*

Thus, 2, 3, 6 are in harmonical proportion; for

$$2 : 6 :: 3-2 : 6-3.$$

Let a, b, c be in harmonical proportion; then

$$a : c :: a-b : b-c.$$

Multiplying the extremes and means, and reducing, we have

$$c = \frac{ab}{2a-b}.$$

Hence, to find a third harmonical proportional to two quantities, divide the product of the first and second by twice the first diminished by the second.

Ex. 1. Find a third harmonical proportional to 3 and 5.

Ex. 2. Find a third harmonical proportional to 5 and 8.

(229.) Four quantities are said to be in harmonical proportion when *the first is to the fourth as the difference between the first and second is to the difference between the third and fourth.*

Thus, 2, 3, 4, 8 are in harmonical proportion; for

$$2 : 8 :: 3-2 : 8-4.$$

Let a, b, c, d be in harmonical proportion; then

$$a : d :: a-b : c-d.$$

Multiplying the extremes and means, and reducing, we have

$$d=\frac{ac}{2a-b}.$$

Hence, to find a fourth harmonical proportional to three quantities, divide the product of the first and third by twice the first diminished by the second.

Ex. 1. Find a fourth harmonical proportional to 4, 5, and 6.

Ex. 2. Find a fourth harmonical proportional to 5, 8, and 10.

(230.) Proportions are often expressed in an *abridged form.* Thus, if A and B represent two sums of money put out for one year at the same rate of interest, then

A : B : : *interest of* A : *interest of* B.

This is briefly expressed by saying that the interest *varies as* the principal. A peculiar character $\propto$ is used to denote this relation. Thus, we write

the interest $\propto$ the principal.

One quantity varies *directly* as another, when both increase or diminish together in the same ratio. Thus, in the above example, A varies directly as the interest of A. In such a case either quantity is equal to the other multiplied by some constant number. Thus, if the interest *varies* as the principal, then the interest *equals* the principal multiplied by a constant quantity, which is the *rate* of interest.

If $\quad A \propto B$, then $A=mB$.

If the space (S) described by a falling body varies as the square of the time (T), then

$$S=mT^2,$$

m representing some constant quantity.

(231.) One quantity may vary directly as the product of several others. Thus, if a body moves with uniform velocity the space described is measured by the product of the time by the velocity. If we put S to represent the space described, T the time of motion, and V the uniform velocity, then we shall have

$$S \propto T \times V.$$

Also the area of a rectangle varies as the product of its length and breadth.

The weight of a stick of timber varies as its length $\times$ its breadth $\times$ its depth $\times$ its density.

If the density is given, then the weight varies as the length × the breadth × the depth.

If the depth also is given, then the weight varies as the length × the breadth.

If the breadth is given, then the weight varies as the length.

Finally, if the length also is given, then the weight is equal to a *constant* quantity.

(232.) One quantity varies *inversely* as another when one increases in the same ratio that the other diminishes. Thus, the altitude of a triangle whose area is given, varies inversely as its base.

If the product of two quantities is constant, then one varies inversely as the other.

In uniform motion, the space is measured by the product of the time by the velocity; that is,

$$S=T\times V.$$

Whence $$T=\frac{S}{V}.$$

If the space be supposed to remain constant, then

$$T \propto \frac{1}{V};$$

that is, the time required to travel a given distance varies inversely as the velocity. Suppose the distance is 360 miles: then,

if the velocity is	12	miles per hour,	the time will be	30	hours;
"	20	"	"	18	"
"	24	"	"	15	"

that is, if the velocity is *doubled*, the time is *halved*. The one varies inversely as the other.

Conversely, if one quantity varies inversely as another, the product of the two quantities is *constant*.

Thus, if $$T \propto \frac{1}{V},$$

then the space (S) is a constant quantity.

(233.) One quantity may vary directly as a second, and inversely as a third. Thus, according to the Newtonian law of gravitation, the attraction (G) of any heavenly body varies directly as the quantity of matter (Q), and inversely as the square of the distance (D).

That is, $$G \propto \frac{Q}{D^2}.$$

(234.) *Application of the preceding principles.*

Ex. 1. Given $x+y : x :: 5 : 3$, $xy=6$, to find the values of x and y.

Since $$x+y : x :: 5 : 3.$$

By division, *Art.* 220, $$y : x :: 2 : 3.$$

Therefore, $$3y=2x, \text{ and } y=\frac{2x}{3}.$$

Substituting this value of y in the second equation, we obtain

$$\frac{2x^2}{3}=6.$$

Therefore, $$x=\pm 3,$$

and $$y=\pm 2.$$

Ex. 2. Given $x+y : x-y :: 3 : 1$, $x^3-y^3=56$, to find the values of x and y.

From the first equation, by *Art.* 222, we obtain

$$2x : 2y :: 4 : 2;$$

whence, $$x : y :: 2 : 1,$$

and $$x=2y.$$

Substituting this value of x in the second equation, we obtain

$$y=2,\ x=4.$$

Ex. 3. Given $\overline{x+y}^2 : \overline{x-y}^2 :: 64 : 1$, $xy=63$, to find the values of x and y.

By *Art.* 223, $$x+y : x-y :: 8 : 1.$$

By *Art.* 222, $$2x : 2y :: 9 : 7;$$

whence $$x : y :: 9 : 7.$$

Therefore, $$x=\frac{9y}{7}.$$

Substituting this value of x in the second equation, we obtain

$$y=\pm 7,\ x=\pm 9.$$

Ex. 4. Given $x^3-y^3 : \overline{x-y}^3 :: 61 : 1,$ $xy=320,$ to find the values of x and y.

Since $x^3-y^3 : x^3-3x^2y+3xy^2-y^3 :: 61 : 1.$

By division, *Art.* 220, $3xy\times(x-y) : \overline{x-y}^3 :: 60 : 1.$

Hence $960 : \overline{x-y}^2 :: 60 : 1,$

and $16 : \overline{x-y}^2 :: 1 : 1.$

Therefore, $x-y=\pm4.$

Also, since $x^2-2xy+y^2=16.$

And $4xy=1280.$

By addition, $x^2+2xy+y^2=1296.$

Extracting the root, $x+y=\pm36.$

Hence $x=\pm20, \text{ or } \pm16,$
$y=\pm16, \text{ or } \pm20.$

Ex. 5. Given $x^3-y^3 : x^2y-xy^2 :: 7 : 2,$ $x+y=6,$ to find the values of x and y.

Ans. $x=4$, or 2; $y=2$, or 4.

Ex. 6. Given $\sqrt{y} - \sqrt{a-x} = \sqrt{y-x},$ $\sqrt{y-x}+\sqrt{a-x} : \sqrt{a-x} :: 5 : 2,$ to find the values of x and y.

Ans. $x=\frac{4a}{5}$; $y=\frac{5a}{4}$.

Ex. 7. Given $x+\sqrt{x} : x-\sqrt{x} :: 3\sqrt{x}+6 : 2\sqrt{x}$, to find the values of x.

Ans. $x=9$, or 4.

Ex. 8. What number is that to which, if 1, 5, and 13 be severally added, the first sum shall be to the second as the second to the third?

Ans. 3.

Ex. 9. What number is that to which, if a, b, and c be severally added, the first sum shall be to the second as the second to the third?

Ans. $\frac{b^2-ac}{a-2b+c}$.

Ex. 10. What two numbers are those whose difference, sum, and product are as the numbers 2, 3, and 5 respectively?

Ans. 2 and 10.

Ex. 11. What two numbers are those whose difference, sum, and product are as the numbers m, n, and p?

Ans. $\frac{2p}{n+m}$, and $\frac{2p}{n-m}$

Ex. 12. Find two numbers, the greater of which shall be to the less as their sum to 42, and as their difference to 6.

Ans. 32 and 24.

Ex. 13. Find two numbers, the greater of which shall be to the less as their sum to a, and their difference to b.

Ans. $\frac{(a+b)^2}{2(a-b)}$, and $\frac{a+b}{2}$.

Ex. 14. There are two numbers which are in the ratio of 3 to 2, the difference of whose fourth powers is to the sum of their cubes as 26 to 7. Required the numbers.

Ans. 6 and 4.

Ex. 15. What two numbers are in the ratio of m to n, the difference of whose fourth powers is to the sum of their cubes as p to q?

Ans. $\frac{mp}{q}\times\frac{m^3+n^3}{m^4-n^4}$, and $\frac{np}{q}\times\frac{m^3+n^3}{m^4-n^4}$.

SECTION XIV.

PROGRESSIONS.

ARITHMETICAL PROGRESSION.

(235.) *An Arithmetical Progression is a series of quantities which increase or decrease by the continued addition or subtraction of the same quantity.*

Thus, the numbers

$$1, 3, 5, 7, 9, 11, \&c.,$$

which are obtained by the addition of 2 to each successive term, form what is called an *increasing* Arithmetical Progression; and the numbers

$$20, 17, 14, 11, 8, 5, \&c.,$$

which are obtained by the subtraction of 3 from each successive term, form what is called a *decreasing* Arithmetical Progression.

(236.) *To find the last term of an Arithmetical Progression.*

If a represent the first term of an arithmetical progression, and d the common difference, the successive terms of an increasing series will be

$$a, a+d, a+2d, a+3d, a+4d, \&c.$$

The successive terms of a decreasing series will be

$$a, a-d, a-2d, a-3d, a-4d, \&c.$$

Since the coefficient of d in the *second* term is 1, in the *third* term 2, in the *fourth* term 3, and so on, the nth term of the series will be

$$a \pm (n-1)d,$$

which may be called the *last* term when the number of terms is n. Hence,

The last term of an arithmetical progression is equal to the first, ± the product of the common difference into the number of terms less one.

In what follows we shall consider the progression an *increasing* one, since all the results which we obtain can be immediately applied to a decreasing series by changing the sign of d.

If we put l to represent the last term of the series, we shall accordingly have

$$l=a+(n-1)d.$$

This equation contains four variable quantities, any one of which may be computed when the other three are known.

(237.) *To find the sum of* n *terms of the series.*

Take any series, and under it set the same terms in an inverted order, thus:

Let the series be	1, 3, 5, 7, 9, 11, 13, 15,
the same series inverted is	15, 13, 11, 9, 7, 5, 3, 1.
The sums are,	16, 16, 16, 16, 16, 16, 16, 16.

The sums of the two series must be double the sum of a single series, and is equal to the sum of the extremes repeated as many times as there are terms.

In order to generalize this method, let S represent the sum of the series,

Then $S=a+\overline{a+d}+\overline{a+2d}+\overline{a+3d}+\ldots\ldots\ldots\ldots+l.$

If we write the same series in an inverted order, thus:

$$S=l+\overline{l-d}+\overline{l-2d}+\overline{l-3d}+\ldots\ldots\ldots\ldots+a,$$

and add the two series together, term by term, we obtain

$$2S=\overline{l+a}+\overline{l+a}+\overline{l+a}+\overline{l+a}+\ \ldots\ldots\ldots+\overline{l+a}.$$

Represent the number of terms in the series by n; then

$$2S=n(l+a).$$

Hence $$S=\frac{n(l+a)}{2}.$$

Therefore,

The sum of an Arithmetical Progression is equal to half the sum of the two extremes, multiplied by the number of terms.

It also appears from the above, that *the sum of the extremes is equal to the sum of any other two terms equally distant from the extremes.*

(238.) The two fundamental equations

$$l=a+(n-1)d,$$

$$S=\frac{l+a}{2}\cdot n,$$

contain five variable quantities,

$$a, l, d, n, S,$$

of which any three being given, the other two may be found. Accordingly, 20 different cases may arise, all of which are solved by combining the formulæ above given. These cases are exhibited in the following table, and should be verified by the student:

No.	Given.	Required.	Formulæ.
1	a, d, n		$l=a+(n-1)d,$
2	a, d, S		$l=-\frac{1}{2}d\pm\sqrt{2dS+(a-\frac{1}{2}d)^2},$
3	a, n, S	l	$l=\frac{2S}{n}-a,$
4	d, n, S		$l=\frac{S}{n}+\frac{(n-1)d}{2}.$
5	a, d, n		$S=\frac{1}{2}n\{2a+(n-1)d\},$
6	a, d, l		$S=\frac{l+a}{2}+\frac{l^2-a^2}{2d},$
7	a, n, l	S	$S=\frac{l+a}{2}\times n,$
8	d, n, l		$S=\frac{1}{2}n\{2l-(n-1)d\}.$
9	a, n, l		$d=\frac{l-a}{n-1},$
10	a, n, S		$d=\frac{2S-2an}{n(n-1)},$
11	a, l, S	d	$d=\frac{l^2-a^2}{2S-l-a},$
12	n, l, S		$d=\frac{2nl-2S}{n(n-1)}.$
13	d, n, l		$a=l-(n-1)d,$
14	d, n, S		$a=\frac{S}{n}-\frac{(n-1)d}{2},$
15	d, l, S	a	$a=\frac{1}{2}d\pm\sqrt{(l+\frac{1}{2}d)^2-2dS},$
16	n, l, S		$a=\frac{2S}{n}-l.$

No.	Given.	Required.	Formulæ.
17	a, d, l		$n=\frac{l-a}{d}+1,$
18	a, d, S		$n=\frac{\pm\sqrt{(2a-d)^2+8dS}-2a+d}{2d},$
19	a, l, S	n	$n=\frac{2S}{l+a},$
20	d, l, S		$n=\frac{2l+d\pm\sqrt{(2l+d)^2-8dS}}{2d}.$

EXAMPLES.

(239.) *Ex.* 1. Required the sum of 60 terms of an arithmetical progression whose first term is 5, and common difference 10.

Ans. 18000.

This example affords an application of Formula 5.

Ex. 2. Required the number of terms of a progression whose sum is 442, whose first term is 2, and common difference 3.

Ans. 17.

This example is solved by Formula 18.

Ex. 3. Required the first term of a progression whose sum is 99, whose last term is 19, and common difference 2.

Ans. 3.

Ex. 4. The sum of a progression is 1455, the first term 5, and the last term 92. What is the common difference?

Ans. 3.

Ex. 5. A body falls 16 feet during the first second, and in each succeeding second 32 feet more than in the one immediately preceding. If it continue falling for 20 seconds, how many feet will it pass over in the last second, and how many in the whole time?

Ans. 624 feet in the last second, and 6400 feet in the whole time.

Ex. 6. Required the sum of 101 terms of the series

1, 3, 5, 7, 9, &c.

Ans. 10201.

Ex. 7. Find the nth term of the series

1, 3, 5, 7, 9, &c.

Ans. $2n-1$;

that is, *the last term of this series is one less than twice the number of terms.*

Ex. 8. Find the sum of n terms of the series

1, 3, 5, 7, 9, &c.

Ans. n^2;

that is, *the sum of the terms of this series is equal to the square of the number of terms.*

Thus,
$$1+3 \qquad\quad = 4=2^2.$$
$$1+3+5 \qquad = 9=3^2.$$
$$1+3+5+7 \quad =16=4^2.$$
$$1+3+5+7+9=25=5^2.$$

Ex. 9. Find the sum of the natural series of numbers

1, 2, 3, 4, 5, &c.,

up to n terms.

Ans. $\frac{n(n+1)}{2}$.

Ex. 10. Find the sum of the even numbers

2, 4, 6, 8, &c.,

up to n terms.

Ans. $n(n+1)$.

Ex. 11. One hundred stones being placed on the ground in a straight line, at the distance of two yards from each other, how far will a person travel who shall bring them one by one to a basket which is placed two yards from the first stone?

Ans. 20200 yards.

Ex. 12. Find m arithmetical means between two given numbers.

In order to solve this problem, we must first find the common difference. The whole number of terms consists of the two extremes and all the intermediate terms. If, then, m represent the number of means, $m+2$ will be the whole number of terms.

Substituting $m+2$ for n, in Formula 9, page 190, we have

$$d=\frac{l-a}{m+1}=\text{ the common difference,}$$

whence the required means are easily obtained by addition.

Ex. 13. Find 6 arithmetical means between 1 and 50.

Ex. 14. Find three numbers in arithmetical progression, the sum of whose squares shall be 1232, and the square of the mean greater than the product of the two extremes by 16.

Ans. 16, 20, and 24.

In examples of this kind, it is generally best to represent the series in such a manner that the common difference may disappear in taking the sum of the terms. Thus a progression of three terms may be represented by

$$a-d,\ a,\ a+d;$$

one of four terms by $a-3d$, $a-d$, $a+d$, $a+3d$, &c.

Ex. 15. Find three numbers in arithmetical progression, the sum of whose squares shall be a, and the square of the mean greater than the product of the two extremes by b.

Ans. $\sqrt{\frac{a-2b}{3}}-\sqrt{b}$; $\sqrt{\frac{a-2b}{3}}$; and $\sqrt{\frac{a-2b}{3}}+\sqrt{b}$.

Ex. 16. Find four numbers in arithmetical progression whose sum is 28, and continued product 585.

Ans. 1, 5, 9, 13.

Ex. 17. A sets out for a certain place, and travels 1 mile the first day, 2 the second, 3 the third, and so on. In five days afterward B sets out, and travels 12 miles a day. How long will A travel before he is overtaken by B?

Ans. 8 or 15 days.

This is another example of an equation of the second degree, in which the two roots are both positive. The following diagram exhibits the daily progress of each traveler. The divisions above the horizontal line represent the distances traveled each day by A; those below the line the distances traveled by B.

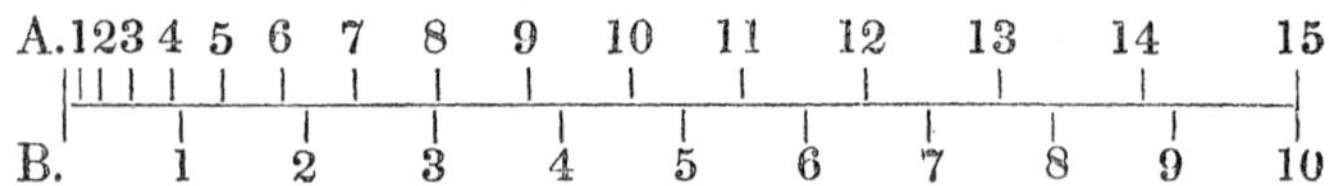

It is readily seen from the figure that A is in advance of B until the end of his 8th day, when B overtakes and passes him. After the 12th day, A gains upon B, and passes him on the 15th day, after which he is continually gaining upon B, and could not be again overtaken.

Ex. 18. A goes 1 mile the first day, 2 the second, and so on.

B starts a days later, and travels b miles per day. How long will A travel before he is overtaken by B?

Ans. $\dfrac{2b-1\pm\sqrt{(2b-1)^2-8ab}}{2}$ days.

In what case would B *never* overtake A?

Ans. When $a>\dfrac{(2b-1)^2}{8b}$.

For instance, in the preceding example, if B had started one day later, he could never have overtaken A.

Ex. 19. A traveler set out from a certain place and went 1 mile the first day, 3 the second, 5 the third, and so on. After he had been gone three days, a second traveler sets out, and goes 12 miles the first day, 13 the second, and so on. In how many days will the second overtake the first?

Ans. In 2 or 9 days.

Let the student illustrate this example by a diagram like the preceding.

GEOMETRICAL PROGRESSION.

(240.) *A Geometrical Progression is a series of quantities, each of which is equal to the product of that which precedes it by a constant number.*

Thus, the series

$$2,\ 4,\ 8,\ 16,\ 32,\ \&c.,$$

and

$$81,\ 27,\ 9,\ 3,\ \&c.,$$

are geometrical progressions. In the former, each number is derived from the preceding by multiplying it by 2, and the series forms an *increasing* geometrical progression. In the latter, each number is derived from the preceding by multiplying it by $\frac{1}{3}$, and the series forms a *decreasing* geometrical progression.

In each of these cases, the common multiplier is called the *common ratio.*

(241.) *To find the last term of a geometrical progression.*

Let a represent the first term of the progression, and r the common ratio; then the successive terms of the series will be

$$a,\ ar,\ ar^2,\ ar^3,\ ar^4,\ \&c.$$

The exponent of r in the *second* term is 1, in the *third* term is 2, in the *fourth* term 3, and so on; hence the nth term of the series will be

$$ar^{n-1}.$$

If, therefore, we put l for the last term, and n the number of terms of the series, we shall have

$$l=ar^{n-1}.$$

That is,

The last term of a geometrical progression is equal to the product of the first term by that power of the ratio whose exponent is one less than the number of terms.

(242.) *To find the sum of all the terms of a geometrical progression.*

If we take any geometrical series, and multiply each of its terms by the ratio, a new series will be formed, of which every term except the last will have its corresponding term in the first series. Thus, take the series

$$1, 2, 4, 8, 16, 32,$$

the sum of which we will represent by S, so that

$$S=1+2+4+8+16+32.$$

Multiplying each term by 2, we obtain

$$2S=2+4+8+16+32+64.$$

The terms of the two series are identical, except the *first* term of the first series and the *last* term of the second series. If, then, we subtract one of these equations from the other, all the remaining terms will disappear, and we shall have

$$2S-S=64-1.$$

In order to generalize this method, let a, ar, ar^2, &c., represent any geometrical series, and S its sum; then

$$S=a+ar+ar^2+ar^3+\ldots\ldots+ar^{n-2}+ar^{n-1}.$$

Multiplying this equation by r, we have

$$rS=ar+ar^2+ar^3+ar^4+\ldots\ldots+ar^{n-1}+ar^n.$$

Subtracting the first equation from the second, we obtain

$$rS-S=ar^n-a.$$

Hence $$S=\frac{ar^n-a}{r-1};$$

or, substituting the value of l already found, we shall have

$$S=\frac{lr-a}{r-1}.$$

Hence, to find the sum of the terms of a geometrical progression,

Multiply the last term by the ratio, subtract the first term, and divide the remainder by the ratio less one.

If the series is a decreasing one, and r consequently represents a fraction, it is convenient to change the signs of both numerator and denominator in this expression, which then becomes

$$S=\frac{a-ar^n}{1-r}=\frac{a-lr}{1-r}.$$

(243.) In the two fundamental equations

$$l=ar^{n-1},$$
$$S=\frac{lr-a}{r-1},$$

there are five variable quantities,

$$a,\ l,\ r,\ n,\ S,$$

of which *any three* being given, the other two may be found. Accordingly, as in arithmetical progression, 20 different cases may arise, all of which are readily solved, with the exception of those in which n is the quantity sought. The value of n can only be found by the solution of an exponential equation. See *Art.* 352. These different cases are all exhibited in the following table for convenient reference.

No.	Given.	Required.	Formulæ.
1	a, r, n		$l=ar^{n-1}$,
2	a, r, S		$l=\frac{a+(r-1)S}{r}$,
3	a, n, S	l	$l(S-l)^{n-1}=a(S-a)^{n-1}$,
4	r, n, S		$l=\frac{(r-1)Sr^{n-1}}{r^n-1}$.
5	a, r, n		$S=\frac{ar^n-a}{r-1}$,
6	a, r, l		$S=\frac{lr-a}{r-1}$,
7	a, n, l	S	$S=\frac{\sqrt[n-1]{l^n}-\sqrt[n-1]{a^n}}{\sqrt[n-1]{l}-\sqrt[n-1]{a}}$,
8	r, n, l		$S=\frac{lr^n-l}{r^n-r^{n-1}}$.

No.	Given.	Required.	Formulæ.
9	a, n, l	r	$r=\sqrt[n-1]{\frac{l}{a}}$,
10	a, n, S		$ar^n-rS=a-S$.
11	a, l, S		$r=\frac{S-a}{S-l}$,
12	n, l, S		$(S-l)r^n-Sr^{n-1}=-l$.
13	r, n, l	a	$a=\frac{l}{r^{n-1}}$,
14	r, n, S		$a=\frac{(r-1)S}{r^n-1}$,
15	r, l, S		$a=lr-(r-1)S$,
16	n, l, S		$a(S-a)^{n-1}=l(S-l)^{n-1}$.
17	a, r, l	n	$n=\frac{log.\ l-log.\ a}{log.\ r}+1$,
18	a, r, S		$n=\frac{log.[a+(r-1)S]-log.\ a}{log.\ r}$,
19	a, l, S		$n=\frac{log.\ l-log.\ a}{log.\ (S-a)-log.\ (S-l)}+1$,
20	r, l, S		$n=\frac{log.\ l-log.\ [lr-(r-1)S]}{log.\ r}+1$.

EXAMPLES.

Ex. 1. Required the sum of the series

1, 3, 9, 27, &c.,

continued to 12 terms.

Ans. 265720.

This example is solved by Formula 5.

Ex. 2. Required the sum of the series

1, 2, 4, 8, 16, &c.,

continued to 14 terms.

Ans. 16383.

Ex. 3. Given the first term 2, the ratio 3, and the number of terms 10, to find the last term.

Ans. 39366.

Ex. 4. Given the first term 1, the last term 512, and the sum of the terms 1023, to find the ratio.

Ex. 5. Given the last term 2048, the number of terms 12, and the ratio 2, to find the first term.

Ex. 6. A person being asked to dispose of his horse, said he would sell him on condition of receiving one cent for the first nail in his shoes, two cents for the second, and so on, doubling the price of every nail to 32, the number of nails in his four shoes. What would the horse cost at that rate?

Ans. \$42,949,672.95.

(244.) *To find any number of geometrical means between two given numbers.*

In order to solve this problem, it is necessary to know the *ratio*. If m represent the number of *means*, $m+2$ will be the whole number of *terms*. Substituting $m+2$ for n in Formula 9 *Art.* 243, we obtain

$$r=\sqrt[m+1]{\frac{l}{a}}.$$

That is, to find the ratio, *divide the last term by the first term, and extract the root denoted by the number of means plus one.*

When the ratio is known, the required means are obtained by continued multiplication.

Ex. 1. Find three geometrical means between 2 and 162.

Ex. 2. Find two geometrical means between 4 and 256.

(245.) *Of decreasing progressions having an infinite number of terms.*

The formula

$$S=\frac{a-ar^n}{1-r},$$

which represents the sum of n terms of a decreasing series, may be put under the form

$$S=\frac{a}{1-r}-\frac{ar^n}{1-r}.$$

In a decreasing progression, since r is a proper fraction, r^n is less than unity, and the larger the number n, the smaller will be the quantity r^n. If, therefore, we take a very large number of terms of the series, the quantity r^n, and, consequently the term $\frac{ar^n}{1-r}$, will be very small; and if we take n greater than any assignable number, then $\frac{ar^n}{1-r}$ will be less than any assignable number. We shall therefore have

$$S=\frac{a}{1-r}.$$

Hence the sum of an infinite series decreasing in geometrical progression is found by the following

RULE.

Divide the first term by unity diminished by the ratio.

Ex. 1. Find the sum of the infinite series

$$1+\tfrac{1}{2}+\tfrac{1}{4}+\tfrac{1}{8}+, \&c.$$

Here $a=1,\ r=\tfrac{1}{2}.$

Therefore, $S=\frac{a}{1-r}=\frac{1}{1-\frac{1}{2}}=2.$

Ex. 2. Find the sum of the infinite series

$$1+\tfrac{1}{3}+\tfrac{1}{9}+\tfrac{1}{27}+, \&c.$$

Ans. $\tfrac{3}{2}$.

Ex. 3. Find the sum of the infinite series

$$1+\tfrac{1}{4}+\tfrac{1}{16}+\tfrac{1}{64}+, \&c.$$

Ex. 4. Find the ratio of an infinite progression, whose first term is 1, and the sum of the series $\tfrac{5}{4}$.

Ans. $\tfrac{1}{5}$.

Ex. 5. Find the first term of an infinite progression, whose ratio is $\tfrac{1}{10}$, and the sum $\tfrac{2}{3}$.

Ans. $\tfrac{3}{5}$

Ex. 6. Find the first term of an infinite progression, of which the ratio is $\frac{1}{n}$, and the sum $\frac{n}{n-1}$.

PROBLEMS.

(246.) Prob. 1. Of four numbers in geometrical progression, the sum of the first and second is 15, and the sum of the third and fourth is 60. Required the numbers.

Let $x,\ xy,\ xy^2,\ xy^3$, be the numbers.

Therefore, $x+xy=15,$

and $xy^2+xy^3=60.$

Multiplying the first equation by y^2,

$$xy^2+xy^3=15y^2=60.$$

Therefore, $y^2=4$,

and $y=\pm 2$.

Also, $x\pm 2x=15$.

Therefore, $x=5$, or -15.

Taking the first value of x, and the corresponding value of y, we obtain the series

$$5, 10, 20, 40;$$

which numbers may be easily verified.

Taking the second value of x, and the corresponding value of y, we obtain the series

$$-15, +30, -60, +120;$$

which numbers also perfectly satisfy the problem understood algebraically. If, however, it is required that the terms of the progression be *positive*, the last value of x would be inapplicable to the problem, though satisfying the algebraic equation.

Several of the following problems also have two solutions, if we admit negative values.

Prob. 2. There are three numbers in geometrical progression whose sum is 210, and the last exceeds the first by 90 What are the numbers?

Ans. 30, 60, and 120.

Prob. 3. There are three numbers in geometrical progression whose continued product is 64, and the sum of their cubes is 584. Required the numbers.

Ans. 2, 4, and 8.

Prob. 4. There are four numbers in geometrical progression, the second of which is less than the fourth by 24; and the sum of the extremes is to the sum of the means as 7 to 3. Required the numbers.

Ans. 1, 3, 9, and 27.

Prob. 5. Of four numbers in geometrical progression, the difference between the first and second is 4, and the difference between the third and fourth is 36. What are the numbers?

Ans. 2, 6, 18, and 54.

Prob. 6. Of four numbers in geometrical progression, the

sum of the first and third is a, the sum of the second and fourth is b. What are the numbers?

$$Ans.\ \frac{a^3}{a^2+b^2},\ \frac{a^2b}{a^2+b^2},\ \frac{ab^2}{a^2+b^2},\ \frac{b^3}{a^2+b^2}.$$

HARMONICAL PROGRESSION.

(247.) *A series of quantities is said to be in harmonical progression when, of any three consecutive terms, the first is to the third as the difference of the first and second is to the difference of the second and third.*

Thus the numbers

$$60,\ 30,\ 20,\ 15,\ 12,\ 10,$$

are in harmonical progression; for

$$60 : 20 :: 60-30 : 30-20$$
$$30 : 15 :: 30-20 : 20-15$$
$$20 : 12 :: 20-15 : 15-12$$
$$15 : 10 :: 15-12 : 12-10.$$

So, also, the numbers

$$1,\ \tfrac{1}{2},\ \tfrac{1}{3},\ \tfrac{1}{4},\ \tfrac{1}{5},\ \tfrac{1}{6},\ \&c.,$$

form an harmonical progression.

(248.) *The reciprocals of a series of terms in harmonical progression form an arithmetical progression.*

Thus, the reciprocals of 60, 30, 20, &c., are

$$\tfrac{1}{60},\ \tfrac{1}{30},\ \tfrac{1}{20},\ \tfrac{1}{15},\ \tfrac{1}{12},\ \tfrac{1}{10},$$

which are respectively equal to

$$\tfrac{1}{60},\ \tfrac{2}{60},\ \tfrac{3}{60},\ \tfrac{4}{60},\ \tfrac{5}{60},\ \tfrac{6}{60},$$

being an arithmetical progression whose common difference is $\frac{1}{60}$.

If six musical strings of equal weight and tension have their lengths in the ratio of the numbers

$$1,\ \tfrac{1}{2},\ \tfrac{1}{3},\ \tfrac{1}{4},\ \tfrac{1}{5},\ \tfrac{1}{6},$$

the second will sound the octave of the first; the third will sound the twelfth; the fourth will sound the double octave; the fifth will sound the eighteenth; and the sixth will sound the third octave of the first. Hence the origin of the term *harmonical* or *musical* proportion.

Let a, b, c be three quantities in harmonical progression; then

$$a : c :: a-b : b-c;$$

whence

$$b=\frac{2ac}{a+c}.$$

That is, *an harmonical mean between two quantities is equal to twice their product divided by their sum.*

SECTION XV.

GREATEST COMMON DIVISOR.—CONTINUED FRACTIONS.—PERMUTATIONS AND COMBINATIONS.

(249.) The greatest common divisor of two or more quantities is the greatest factor which is common to each of the quantities.

THEOREM.

The greatest common divisor of two quantities is the same with the greatest common divisor of the least quantity, and their remainder after division.

To prove this principle, let the greatest of the two quantities be represented by A, and the least by B. Divide A by B; let the entire part of the quotient be represented by Q, and the remainder by R. Then, since the dividend must be equal to the product of the divisor by the quotient + the remainder, we shall have

$$A=QB+R.$$

Now every number which will divide B will divide QB; and every number which will divide R and QB will divide R+QB or A. That is, every number which is a common divisor of B and R is a common divisor of A and B.

Again, every number which will divide A and B will divide A and QB; it will also divide A−QB or R. That is, every number which is a common divisor of A and B is also a common divisor of B and R. Hence the greatest common divisor of A and B must be the same as the greatest common divisor of B and R.

(250.) To find, then, the greatest common divisor of two quantities, we divide the greater by the less; and the remainder, which is necessarily less than either of the given quantities, is by the last Article divisible by the greatest common divisor.

Dividing the preceding divisor by the last remainder, a still smaller remainder will be found, which is divisible by the greatest common divisor; and by continuing this process with each remainder and the preceding divisor, quantities smaller and smaller are found, which are all divisible by the greatest common divisor, until at length the greatest common divisor must be obtained. Hence the following

RULE.

Divide the greater quantity by the less, and the preceding divisor by the last remainder, till nothing remains; the last divisor will be the greatest common divisor.

When the remainders decrease to unity, the given quantities have *no* common divisor greater than unity, and are said to be *incommensurable*, or *prime* to each other.

EXAMPLES.

Ex. 1. What is the greatest common divisor of **372** and **246**?

```
            372|246
            246| 1
        246|126, first Remainder.
        126| 1
    126|120, second Remainder.
    120| 1
120|  6, third Remainder.
120| 20
```

Here we have continued the operation of division until we obtain 0 for a remainder; the last divisor (6) is the greatest common divisor. Thus, 246 and 372 being each divided by 6, give 41 and 62, and these quotients are *prime* with respect to each other; that is, have no common divisor greater than unity.

Ex. 2 What is the greatest common divisor of

336 and 720?

Ans. 48.

Ex. 3. What is the greatest common divisor of

918 and 522?

Ans. 18.

(251.) In applying this rule to polynomials, some modification may become necessary. It may happen that the first term of the dividend is not divisible by the first term of the divisor. This may arise from the presence of a factor in the divisor which is not found in the dividend, and may therefore be suppressed. For, since the greatest common divisor of two quantities is only the product of their *common* factors, it can not be affected by a factor of the one quantity which is *not found* in the other.

We may therefore suppress in the first polynomial all the factors common to each of its terms. We do the same with the second polynomial, and if the suppressed factors have a common divisor, we reserve it as forming part of the common divisor sought.

But if, after this reduction, the first term of the dividend, when arranged according to the powers of some letter, is not divisible by the first term of the arranged divisor, *we may multiply the dividend by any monomial factor which will render its first term divisible by the first term of the divisor.*

This will not affect the greatest common divisor, because we introduce into the dividend a factor which belongs *only* to the first term of the divisor; for by supposition, all the factors common to each of its terms have been suppressed.

EXAMPLES.

Ex. 1. Required the greatest common divisor of

x^5+x^3 and x^4-1.

The operation will here stand as follows:

$$\begin{array}{l|l} x^5+x^3 & x^4-1 \\ \underline{x^5-x} & x \\ \quad\;\; x^3+x, & \text{first Remainder.} \end{array}$$

Suppressing x, we have x^2+1.

$$\begin{array}{l|l} x^4-1 & x^2+1 \\ \underline{x^4+x^2} & \overline{x^2-1} \\ \quad -x^2-1 & \\ \quad -x^2-1. & \end{array}$$

Whence x^2+1 is the greatest common divisor. To verif this result, divide x^5+x^3 by x^2+1, and we obtain x^3; divid x^4-1 by x^2+1, and we obtain x^2-1.

Ex. 2. Required the greatest common divisor of

$$x^3-b^2x \text{ and } x^2+2bx+b^2.$$

Suppressing the factor x in the first polynomial, we procee as follows:

$$\begin{array}{l|l} x^2+2bx+b^2 & x^2-b^2 \\ \underline{x^2 \qquad\quad -b^2} & \overline{1} \\ \qquad 2bx+2b^2, \text{ first Remainder.} & \end{array}$$

Suppressing the factor $2b$,

$$\begin{array}{l|l} x^2-b^2 & x+b \\ \underline{x^2+bx} & \overline{x-b} \\ -bx-b^2 & \\ -bx-b^2 & \end{array}$$

Whence $x+b$ is the greatest common divisor.

Ex. 3. Required the greatest common divisor of

$$4a^3-2a^2-3a+1 \text{ and } 3a^2-2a-1.$$

Ans. $a-1$.

Ex. 4. Find the greatest common divisor of

$$x^4-a^4 \text{ and } x^3-a^3.$$

Ans. $x-a$.

Ex. 5. Find the greatest common divisor of

$$a^2-3ab+2b^2 \text{ and } a^2-ab-2b^2.$$

Ans. $a-2b$.

Ex. 6. Find the greatest common divisor of

$$a^4-x^4 \text{ and } a^3-a^2x-ax^2+x^3.$$

Ans. a^2-x^2

Ex. 7. Find the greatest common divisor of

$$a^3-a^2b+3ab^2-3b^3 \text{ and } a^2-5ab+4b^2.$$

Ans. $a-b$

Ex. 8. Find the greatest common divisor of

$$a^2-3ab+ac+2b^2-2bc \text{ and } a^2-b^2+2bc-c^2.$$

CONTINUED FRACTIONS.

(252.) From the operation on page 204, we see that the fraction $\frac{246}{372}$ is equal to $\frac{1}{1+\frac{126}{246}}$.

Also, the fraction $\frac{126}{246}$ is equal to $\frac{1}{1+\frac{120}{126}}$.

Therefore, $\frac{246}{372}$ is equal to $\cfrac{1}{1+\cfrac{1}{1+\frac{120}{126}}}$.

Again, the fraction $\frac{120}{126}$ is equal to $\frac{1}{1+\frac{6}{120}}$ or $\frac{1}{1+\frac{1}{20}}$.

Therefore, $\frac{246}{372}$ is equal to

$$\cfrac{1}{1+\cfrac{1}{1+\cfrac{1}{1+\frac{1}{20}}}},$$

which is called a *continued fraction.*

A continued fraction is one whose numerator is unity, and its denominator an integer plus a fraction whose numerator is likewise unity, and its denominator an integer plus a fraction, and so on.

The general form of a continued fraction is

$$\cfrac{1}{a+\cfrac{1}{b+\cfrac{1}{c+\cfrac{1}{d+\cfrac{1}{e+1, \&c.}}}}}$$

(253.) Any fraction may be transformed into a continued fraction by the method of finding the greatest common divisor of the numerator and denominator.

Ex. 1. Transform $\frac{114}{347}$ into a continued fraction.

Ans. $\cfrac{1}{3+\cfrac{1}{22+\cfrac{1}{1+\frac{1}{4}}}}$.

Ex. 2. Transform $\frac{351}{965}$ into a continued fraction.

Ans. $$\cfrac{1}{2+\cfrac{1}{1+\cfrac{1}{2+\cfrac{1}{1+\frac{1}{87}}}}}.$$

Ex. 3. Transform $\frac{421}{972}$ into a continued fraction.

Ex. 4. Transform $\frac{251}{764}$ into a continued fraction.

Ex. 5. Transform $\frac{130}{421}$ into a continued fraction.

(254.) The value of a continued fraction, when composed a finite number of terms, is easily found.

Ex. 1. Find the value of the continued fraction

$$\cfrac{1}{2+\cfrac{1}{3+\frac{1}{4}}}.$$

Beginning with the last fraction, we have

$$3+\tfrac{1}{4}=\frac{13}{4}.$$

Hence $$\frac{1}{3+\frac{1}{4}}=\frac{4}{13}.$$

Therefore, $$2+\frac{1}{3+\frac{1}{4}}=\frac{30}{13}.$$

And $$\cfrac{1}{2+\cfrac{1}{3+\frac{1}{4}}}=\frac{13}{30}. \quad \textit{Ans.}$$

Ex. 2. Find the value of the continued fraction

$$\cfrac{1}{3+\cfrac{1}{2+\cfrac{1}{4+\frac{1}{5}}}}.$$

Ex. 3. Find the value of the continued fraction

$$\cfrac{1}{2+\cfrac{1}{3+\cfrac{1}{2+\cfrac{1}{2+\frac{1}{2}.}}}}$$

(255.) When a fraction has been transformed into a continued fraction, its approximate value may be found by taking a few of the first terms of the continued fraction.

Thus, an approximate value of $\frac{114}{347}$ is $\frac{1}{3}$, which is the first term of its continued fraction.

By taking two terms, we obtain $\frac{22}{67}$, which is a nearer approximation; and three terms would give a still more accurate value.

Ex. 1. Find approximate values of the fraction $\frac{532}{1193}$.

Ans. $\frac{1}{2}, \frac{4}{9}, \frac{33}{74}$.

Ex. 2. Find approximate values of the fraction $\frac{115}{424}$.

Ex. 3. Find approximate values of the fraction $\frac{119}{409}$.

(256.) By this method we are enabled to discover the approximate value of a fraction expressed in large numbers; and this principle has some important applications, particularly in Astronomy.

Ex. 4. The ratio of the circumference of a circle to its diameter is 3.1415926. Find approximate values for this ratio.

Ans. $\frac{22}{7}, \frac{333}{106}, \frac{355}{113}$.

Ex. 5. In 87969 years, the Earth makes 277287 conjunctions with Mercury. Find approximate values for the fraction $\frac{87969}{277287}$.

Ans. $\frac{1}{3}, \frac{6}{19}, \frac{7}{22}, \frac{13}{41}, \frac{33}{104}$.

Ex. 6. In 57551 years, the Earth makes 36000 conjunctions with Venus. Find approximate values for the fraction $\frac{57551}{36000}$.

Ans. $\frac{8}{5}, \frac{235}{147}$.

Ex. 7. In 295306 years, the Moon makes 3652422 synodical revolutions. Find an approximate value of the fraction $\frac{295306}{3652422}$.

Ans. $\frac{19}{235}$.

THEORY OF PERMUTATIONS AND COMBINATIONS.

(257.) The different orders in which quantities may be arranged are called their *Permutations.* Thus,

the permutations of the three letters a, b, c, taken *all together*, are . . . . $\begin{cases} a, b, c, \\ a, c, b, \\ b, a, c, \\ b, c, a, \\ c, a, b, \\ c, b, a. \end{cases}$

The permutations of the same letters taken *two and two*, are . . . . $\begin{cases} a, b, \\ a, c, \\ b, a, \\ b, c, \\ c, a, \\ c, b. \end{cases}$

The permutations of the same letters taken singly, or *one by one*, are . . . . $\begin{cases} a, \\ b, \\ c. \end{cases}$

(258.) *To find the number of permutations of* n *letters, taken* m *and* m *together.*

Let $a, b, c, d \ldots\ldots k$, be the n letters.

The number of permutations of n letters taken singly, or one by one, is evidently equal to the number of letters, or to n.

The number of permutations of n letters taken two and two is $n(n-1)$. For if we reserve one of the letters, as a, there will remain $n-1$ letters,

$$b, c, d \ldots\ldots k.$$

Writing a before each of these letters, we shall have

$$ab, ac, ad \ldots\ldots ak;$$

that is, we obtain $n-1$ permutations of the n letters taken two and two, in which *a stands first.* Proceeding in the same manner with b, we shall find $n-1$ permutations of the n letters taken two and two, in which *b stands first;* and so for each of the n letters. Hence the whole number of permutations will be

$$n(n-1).$$

The number of permutations of n letters taken three and three together is

$$n(n-1)\ (n-2).$$

For if we reserve one of the letters, as a, there will remain $n-1$ letters. Now we have found the number of permutations of n letters taken two and two to be $n(n-1)$. Hence the permutations of $n-1$ letters taken two and two must be

$$(n-1)\ (n-2).$$

Writing a before each of these permutations, we shall have $(n-1)\ (n-2)$ permutations of the n letters taken three and three, in which *a stands first.* Proceeding in the same manner with b, we shall find $(n-1)\ (n-2)$ permutations of the n letters taken three and three, in which *b stands first;* and so for each of the n letters. Hence the whole number of permutations will be

$$n(n-1)\ (n-2).$$

In like manner, we can prove that the number of permutations of n letters taken four and four is

$$n(n-1)\ (n-2)\ (n-3).$$

When the letters are taken *two and two,* the last factor in the formula representing the number of permutations is $n-1$. When the letters are taken *three and three,* the last factor is $n-2$. When the letters are taken *four and four,* the last factor is $n-3$.

Hence, when the letters are taken m and m together, the last factor will be $n-(m-1)$ or $n-m+1$; and the number of permutations of n letters taken m and m together will accordingly be

$$n(n-1)\ (n-2)\ (n-3)\ldots\ldots(n-m+1).$$

EXAMPLES.

Ex. 1. Required the number of permutations of the 8 letters a, b, c, d, e, f, g, h, taken 5 and 5 together.

Here $n=8,\ m=5,\ n-m+1=4,$

and the above formula becomes

$$8.7.6.5.4=6720, \textit{ Ans.}$$

Ex. 2. Required the number of permutations of the 26 letters of the alphabet, taken 4 and 4 together.

Ans. 358800.

Ex. 3. Required the number of permutations of 12 letters, taken 6 and 6 together.

Ans. 665280.

(259.) If we suppose that *each permutation comprehends all the* n *letters;* that is, if $m=n$, the preceding formula becomes

$$n(n-1)\ (n-2)\ \ldots\ldots\ 2\times1;$$

or, inverting the order of the factors,

$$1.2.3.4\ \ldots\ldots\ (n-1)n;$$

which expresses the number of permutations of n letters taken all together.

Ex. 1. Required the number of changes which can be rung upon 8 bells.

According to the preceding formula, we have

$$1.2.3.4.5.6.7.8=40320, \textit{ Ans.}$$

Ex. 2. How many permutations may be formed from the letters of the word *Roma?*

Ex. 3. What is the number of permutations which may be formed from the letters composing the word "*virtue?*"

Ex. 4. What is the number of different arrangements which can be made of 12 persons at a dinner-table?

Ans. 479001600.

(260.) The *combinations* of any number of quantities signify the different collections which may be formed of these quantities, without regard to the *order* of their arrangement.

Thus, the three letters a, b, c, taken *all together*, form but *one* combination, abc.

Taken *two and two*, they form three combinations,

$$ab,\ ac,\ bc.$$

(261.) *To find the number of combinations of* n *letters, taken* m *and* m *together.*

The number of combinations of n letters taken separately, or *one by one*, is evidently n.

The number of combinations of n letters taken *two and two*, is $\frac{n(n-1)}{1.2}$.

For the number of permutations of n letters taken two and two is $n(n-1)$; and there are two permutations (ab, ba) corresponding to one combination of two letters. Therefore the number of combinations will be found by dividing the number of permutations by 2.

The number of combinations of n letters taken *three and three* together, is $\frac{n(n-1)(n-2)}{1.2.3}$.

For the number of permutations of n letters taken three and three, is $n(n-1)(n-2)$; and there are 1.2.3 permutations for one combination of three letters. Therefore the number of combinations will be found by dividing the number of permutations by 1.2.3.

In the same manner, we shall find the number of combinations of n letters, taken m and m together, to be

$$\frac{n(n-1)(n-2) \ldots\ldots (n-m+1)}{1.2.3 \quad \ldots\ldots \quad m}.$$

Ex 1. Required the number of combinations of six letters taken three and three together.

Here $\qquad n=6, m=3, n-m+1=4,$

and the formula becomes

$$\frac{6.5.4}{1.2.3}=20.$$

Ex. 2. Required the number of combinations of 8 letters taken 4 and 4.

Ans. 70.

Ex. 3. Required the number of combinations of 10 letters taken 6 and 6.

Ans. 210.

The following table, which is computed by the preceding formula, shows the number of combinations of 1, 2, 3, 4, &c. letters taken singly, or two and two, three and three, &c. An important application of these principles will be seen in the next Section.

Letters.	Singly.	2 and 2.	3 and 3.	4 and 4.	5 and 5.	6 and 6.	7 and 7.	8 and 8.	9 and 9.	10 and 10.
1	1									
2	2	1								
3	3	3	1		Number of combinations.					
4	4	6	4	1						
5	5	10	10	5	1					
6	6	15	20	15	6	1				
7	7	21	35	35	21	7	1			
8	8	28	56	70	56	28	8	1		
9	9	36	84	126	126	84	36	9	1	
10	10	45	120	210	252	210	120	45	10	1

SECTION XVI.

INVOLUTION OF BINOMIALS.

(262.) We have shown, in *Art.* 142, how to obtain any power of a binomial by actual multiplication. We now propose to develop a theorem by which this labor may be greatly abridged.

Taking the binomial $a+b$, its successive powers found by actual multiplication are as follows:

$$(a+b)^1=a+b,$$
$$(a+b)^2=a^2+2ab+b^2,$$
$$(a+b)^3=a^3+3a^2b+3ab^2+b^3$$
$$(a+b)^4=a^4+4a^3b+6a^2b^2+4ab^3+b^4,$$
$$(a+b)^5=a^5+5a^4b+10a^3b^2+10a^2b^3+5ab^4+b^5,$$
$$(a+b)^6=a^6+6a^5b+15a^4b^2+20a^3b^3+15a^2b^4+6ab^5+b^6.$$

The powers of $a-b$, found in the same manner, are as follows:

$$(a-b)^1=a-b,$$
$$(a-b)^2=a^2-2ab+b^2,$$
$$(a-b)^3=a^3-3a^2b+3ab^2-b^3,$$
$$(a-b)^4=a^4-4a^3b+6a^2b^2-4ab^3+b^4,$$
$$(a-b)^5=a^5-5a^4b+10a^3b^2-10a^2b^3+5ab^4-b^5,$$
$$(a-b)^6=a^6-6a^5b+15a^4b^2-20a^3b^3+15a^2b^4-6ab^5+b^6.$$

On comparing the powers of $a+b$ with those of $a-b$, we perceive that they only differ in *the signs of certain terms.* In the powers of $a+b$, all the terms are positive. In the powers of $a-b$, the terms containing the odd powers of b have the sign $-$, while the even powers retain the sign $+$. The reason of this is obvious; for, since $-b$ is the only negative term of the root, the terms of the power can only be rendered negative by b. A term which contains the factor $-b$ an even number of times, will therefore be positive; if it contain it an odd number of times, it must be negative. Hence it appears that it is only necessary to seek for a method of obtaining the powers of $a+b$; for these will become the powers of $a-b$ by simply changing the signs of the alternate terms.

(263.) If we consider the exponents of the preceding powers, we shall find that they follow a very simple law. Thus,

In the square, the exponents . . $\begin{cases}\text{of } a \text{ are } 2, 1, 0,\\ \text{of } b \text{ are } 0, 1, 2.\end{cases}$

In the cube, the exponents . . $\begin{cases}\text{of } a \text{ are } 3, 2, 1, 0,\\ \text{of } b \text{ are } 0, 1, 2, 3.\end{cases}$

In the fourth power, the exponents $\begin{cases}\text{of } a \text{ are } 4, 3, 2, 1, 0,\\ \text{of } b \text{ are } 0, 1, 2, 3, 4.\end{cases}$

&c., &c., &c.

In the first term of each power, a is raised to the required power of the binomial; and in the following terms, the exponents of a continually decrease by unity to 0; while the exponents of b increase by unity from 0 up to the required power of the binomial. It is obvious that this will always be the case, to whatever extent the involution may be carried. Also, the sum of the exponents of a and b in any term is equal to the exponent of the power required. Thus, in the second power, the sum of the exponents of a and b in each term is 2; in the third power it is 3; in the fourth power, 4, &c.

We hence infer, that for the seventh power the terms, without the coefficients, must be

$$a^7,\ a^6b,\ a^5b^2,\ a^4b^3,\ a^3b^4,\ a^2b^5,\ ab^6,\ b^7;$$

and for the nth power,

$$a^n,\ a^{n-1}b,\ a^{n-2}b^2,\ a^{n-3}b^3 \ .\ .\ .\ .\ .\ a^2b^{n-2},\ ab^{n-1},\ b^n.$$

(264.) It remains to determine the coefficients which belong to these terms; and in order to discover the law of their formation, let us take the coefficients already found by themselves.

The coefficients of the	1st	power are	1 1
"	2d	"	1 2 1
"	3d	"	1 3 3 1
"	4th	"	1 4 6 4 1
"	5th	"	1 5 10 10 5 1
"	6th	"	1 6 15 20 15 6 1

The numbers in this table are identical with those in the table of combinations on page 214. For example, the coefficients of the fifth power denote the number of combinations of five letters taken one and one, two and two, &c.; the coefficients of the sixth power denote the number of combinations of six letters taken one and one, two and two, &c. The reason of this will appear if we observe the law of the product of several binomial factors, $x+a$, $x+b$, $x+c$, $x+d$, &c.

Multiplying $x + a$
by $x + b$,
we obtain $x^2+(a+b)x+ab=$ 1st product.

Multiplying by $x + c$,
we obtain $x^3+(a+b+c)x^2+(ab+ac+bc)x+abc=$ 2d product.

Multiplying by $x + d$,
we obtain $x^4+(a+b+c+d)x^3+(ab+ac+ad+bc+bd+cd)x^2+(abc+abd+acd+bcd)x+abcd=$ 3d product.

We observe that in each of these products *the coefficient of* x *in the first term is unity; the coefficient of the second term is the sum of the second terms of the binomial factors; the coefficient of the third term is the sum of all their products taken two and two; the coefficient of the fourth term is the sum of all their products taken three and three*, &c.

It is easily seen that if we multiply the last product by a new factor, $x+e$, the same law of the coefficients will be preserved. Hence the law is general.

If now, in the preceding binomial factors, we suppose a, b, c, d, &c., to be all equal to each other, the product

$$(x+a)\ (x+b)\ (x+c)\ (x+d)\ \ldots\ldots$$

becomes

$$(x+a)^n.$$

The coefficient of the second term of the product, or $a+b+c+d\ldots\ldots$, becomes $a+a+a+a\ldots\ldots$; that is, a taken as many times as there are letters a, b, c, d, and is, consequently, equal to na.

The coefficient of the third term, or $ab+ac$, &c., reduces to $a^2+a^2+a^2\ldots\ldots$, or a^2 repeated as many times as there are different combinations of n letters taken two and two; that is, by *Art.* 261, to $\frac{n(n-1)}{1.2}a^2$.

The coefficient of the fourth term reduces to a^3 repeated as many times as there are different combinations of n letters taken three and three; that is, $\frac{n(n-1)\,(n-2)}{1.2.3}a^3$, and so on.

Thus we find that the nth power of $x+a$ may be expressed as follows:

$$(x+a)^n=x^n+nax^{n-1}+\frac{n(n-1)}{1.2}a^2x^{n-2}+\frac{n(n-1)\,(n-2)}{1.2.3}a^3x^{n-3}+\ldots$$
$$+na^{n-1}x+a^n,$$

which is called the BINOMIAL FORMULA, and is generally ascribed to Sir Isaac Newton. So important was it regarded, that it was engraved on his monument in Westminster Abbey as one of his greatest discoveries.

On comparing the different terms of this development, we perceive that any coefficient may be derived from the preceding one by the following rule: *If the coefficient of any term be multiplied by the exponent of* x *in that term, and divided by the exponent of* a *increased by one, it will give the coefficient of the succeeding term.*

Thus, the fifth power of $x+a$ is

$$x^5+5ax^4+10a^2x^3+10a^3x^2+5a^4x+a^5.$$

If the coefficient 5 of the second term be multiplied by 4, the exponent of x in that term, and divided by 2, which is the exponent of a increased by one, we obtain 10, the coefficient of the third term.

So, also, if 10, the coefficient of the fourth term, be multiplied by 2, the exponent of x, and divided by 4, the exponent of a increased by one, we obtain 5, the coefficient of the fifth term; and so of the others.

The coefficients of the sixth power will also be found as follows:

$$1,\ 6,\ \frac{6\times5}{2},\ \frac{15\times4}{3},\ \frac{20\times3}{4},\ \frac{15\times2}{5},\ \frac{6\times1}{6};$$

that is, 1, 6, 15, 20, 15, 6, 1.

The coefficients of the seventh power will be

$$1,\ 7,\ \frac{7\times6}{2},\ \frac{21\times5}{3},\ \frac{35\times4}{4},\ \frac{35\times3}{5},\ \frac{21\times2}{6},\ \frac{7\times1}{7};$$

that is, 1, 7, 21, 35, 35, 21, 7, 1.

Therefore, the seventh power of $x+a$ is

$$x^7+7ax^6+21a^2x^5+35a^3x^4+35a^4x^3+21a^5x^2+7a^6x+a^7.$$

It is sometimes preferable to retain the factors of the coefficients distinct from each other, as follows:

$$(x+a)^7=x^7+\frac{7}{1}ax^6+\frac{7.6}{1.2}a^2x^5+\frac{7.6.5}{1.2.3}a^3x^4+\frac{7.6.5.4}{1.2.3.4}a^4x^3+$$
$$\frac{7.6.5.4.3}{1.2.3.4.5}a^5x^2+\frac{7.6.5.4.3.2}{1.2.3.4.5.6}a^6x+\frac{7.6.5.4.3.2.1}{1.2.3.4.5.6.7}a^7.$$

The factor 1 is retained for the sake of symmetry, and to exhibit more clearly the law of the coefficients.

(265.) The following, therefore, is the

BINOMIAL THEOREM.

In any power of a binomial x+a, *the exponent of* x *begins in the first term with the exponent of the power, and in the following terms continually decreases by one. The exponent of* a *commences with one in the second term of the power, and continually increases by one.*

The coefficient of the first term is one; that of the second is the exponent of the power; and if the coefficient of any term be multiplied by the exponent of x *in that term, and divided by the exponent of* a *increased by one, it will give the coefficient of the succeeding term.*

(266.) The *number of terms* in the power is always greater by unity than the exponent of the power. Thus, the number of terms in $(a+b)^4$ is 4+1, or 5; in $(a+b)^6$ is 6+1, or 7.

Also, if we examine the table in *Art.* 264, it will be perceived that, after we pass the middle term, the same coefficients are repeated in the inverse order. Thus, the coefficients of

$(a+b)^5$ are 1, 5, 10, 10, 5, 1;
of $(a+b)^6$ are 1, 6, 15, 20, 15, 6, 1.

Hence it is only necessary to compute the coefficients for *half* the terms; we then repeat the same numbers in the inverse order.

(267.) The *sum of the coefficients* for each power is equal to the number 2 raised to the same power. For, let $x=1$ and $a=1$ then each term without the coefficients reduces to unity, and the value of the power is simply the sum of the coefficients. Also, in this case, $(x+a)^n$ becomes $(1+1)^n$, or 2^n. Thus the coefficients of the

first power are $1+1=2=2^1$;
second " $1+2+1=4=2^2$;
third " $1+3+3+1=8=2^3$;
fourth " $1+4+6+4+1=16=2^4$,
&c., &c., &c.

EXAMPLES.

Ex. 1. Raise $x+a$ to the 9th power.

The terms without the coefficients are

$$x^9,\ ax^8,\ a^2x^7,\ a^3x^6,\ a^4x^5,\ a^5x^4,\ a^6x^3,\ a^7x^2,\ a^8x,\ a^9.$$

And the coefficients are

$$1,\ 9,\ \frac{9\times8}{2},\ \frac{36\times7}{3},\ \frac{84\times6}{4},\ \frac{126\times5}{5},\ \frac{126\times4}{6},\ \frac{84\times3}{7},\ \frac{36\times2}{8},\ \frac{9\times1}{9},$$

that is,

1, 9, 36, 84, 126, 126, 84, 36, 9, 1,

Prefixing the coefficients, we obtain

$$(x+a)^9=x^9+9ax^8+36a^2x^7+84a^3x^6+126a^4x^5+126a^5x^4+84a^6x^3+ +36a^7x^2+9a^8x+a^9.$$

It should be remembered that, according to *Art.* 266, it is only necessary to compute the coefficients of *half* the terms independently.

Ex. 2. What is the 6th power of $x-a$?

(268.) If the terms of the given binomial are affected with coefficients or exponents, they must be raised to the required powers, according to the principles already established for the involution of monomials.

Ex. 3. Raise $2x+5a^2$ to the fourth power.

For convenience, let us substitute b for $2x$, and c for $5a^2$.

Then $(b+c)^4=b^4+4b^3c+6b^2c^2+4bc^3+c^4$.

Restoring the values of b and c,

The first term will be $(2x)^4 \quad =16x^4$.
The second term " $4(2x)^3\times 5a^2 =4.8.5x^3a^2$.
The third term " $6(2x)^2\times(5a^2)^2=6.4.25x^2a^4$.
The fourth term " $4(2x)\times(5a^2)^3=4.2.125xa^6$.
The fifth term " $(5a^2)^4=625a^8$.

Therefore,

$$(2x+5a^2)^4=16x^4+160x^3a^2+600x^2a^4+1000xa^6+625a^8.$$

Ex. 4. What is the fourth power of $2x^3+4y^2$?

Ex. 5. What is the seventh power of $2a-3b$?

Ans. $128a^7-1344a^6b+6048a^5b^2-15120a^4b^3+22680a^3b^4-20412a^2b^5+10206ab^6-2187b^7$.

Ex. 6. What is the sixth power of a^3+3ab?

Ans. $a^{18}+18a^{16}b+135a^{14}b^2+540a^{12}b^3+1215a^{10}b^4++1458a^8b^5+729a^6b^6$.

Ex. 7. What is the fifth power of $5c^2-4y^2z$?

(269.) By means of the Binomial Theorem we can raise any polynomial to any power.

For example, let it be required to raise $a+b+c$ to the third power.

For convenience, we put $b+c=m$; we then have

$$(a+b+c)^3=(a+m)^3=a^3+3a^2m+3am^2+m^3.$$

Substituting for m its equal, $b+c$, we obtain

$$(a+b+c)^3=a^3+3a^2(b+c)+3a(b+c)^2+(b+c)^3.$$

We must now develop the powers of the binomial $b+c$, and perform the multiplications which are here indicated. We thus obtain

$$\begin{aligned}(a+b+c)^3=a^3+3a^2b&+3ab^2+b^3,\\ +3a^2c&+6abc+3b^2c,\\ &+3ac^2+3bc^2,\\ &+c^3.\end{aligned}$$

Ex. 2. Raise $x+a+b$ to the fifth power.

(270.) When one of the terms of a binomial is unity, the powers assume a simpler form, since every power of 1 is 1.

Thus, the fourth power of $a+b$, which is

$$a^4+4a^3b+6a^2b^2+4ab^3+b^4,$$

when we make $a=1$, becomes

$$1+4b+6b^2+4b^3+b^4.$$

So, also, $(1+a)^n=1+\frac{n}{1}a+\frac{n(n-1)}{1.2}a^2+\frac{n(n-1)(n-2)}{1.2.3}a^3+$, &c.

Every binomial of the form $(x+a)^n$ may be reduced to the form of $x^n\left(1+\frac{a}{x}\right)^n$. For

$$x+a \;=x\left(1+\frac{a}{x}\right).$$

Therefore, $$(x+a)^n=x^n\left(1+\frac{a}{x}\right)^n,$$

$$=x^n\left\{1+n.\frac{a}{x}+\frac{n(n-1)}{1.2}.\frac{a^2}{x^2}+\frac{n(n-1)\,(n-2)}{1.2.3}\,\frac{a^3}{x^3}+,\ \&c.\right\}$$

This expression for the value of $(x+a)^n$ is equivalent to that on page 218, as may be easily shown by multiplying x^n into each term within the parenthesis. For some purposes this is regarded as the simplest form.

(271.) In the development of the binomial $(x+a)^n$, we have hitherto supposed n to be a *positive integer*. The Binomial Theorem is, however, applicable, whatever be the nature of the quantity n, whether it be *positive or negative, integral or fractional*. When n is a positive integer, the series consists of $n+1$ terms. In every other case, the series never terminates; that is, the development of $(x+a)^n$ furnishes an *infinite series*.

Ex. 1. It is required to convert $\frac{1}{a+b}$ or $(a+b)^{-1}$ into an infinite series.

According to *Art.* 265, the terms without the coefficients are

$$a^{-1},\ a^{-2}b,\ a^{-3}b^2,\ a^{-4}b^3,\ a^{-5}b^4,\ a^{-6}b^5,\ a^{-7}b^6,\ \&c.$$

The coefficient of the first term is 1.

The coefficient of the second term is -1, the exponent of the power.

The coefficient of the third term is $\frac{-1\times-2}{2}=+1.$

" fourth " $\frac{+1\times-3}{3}=-1.$

" fifth " $\frac{-1\times-4}{4}=+1.$

" sixth " $\frac{+1\times-5}{5}=-1.$

We thus obtain

$$\frac{1}{a+b}=(a+b)^{-1}=a^{-1}-a^{-2}b+a^{-3}b^2-a^{-4}b^3+a^{-5}b^4-a^{-6}b^5+,\ \&c.,$$

where the law of the series is obvious; the coefficients are all unity, and the signs are alternately positive and negative.

We might have obtained the same result by the ordinary method of division. The operation is as follows:

$$\begin{array}{l|l} 1 & a+b \\ \cline{2-2} 1+\frac{b}{a} & \frac{1}{a}-\frac{b}{a^2}+\frac{b^2}{a^3}-\frac{b^3}{a^4}+\text{, \&c.,} = \text{the quotient.} \\ \cline{1-1} \quad -\frac{b}{a} & \\ \quad -\frac{b}{a}-\frac{b^2}{a^2} & \\ \cline{1-1} \qquad +\frac{b^2}{a^2} & \\ \qquad +\frac{b^2}{a^2}+\frac{b^3}{a^3} & \\ \cline{1-1} \qquad\quad -\frac{b^3}{a^3} & \end{array}$$

Hence,

$\frac{1}{a+b}=\frac{1}{a}-\frac{b}{a^2}+\frac{b^2}{a^3}-\frac{b^3}{a^4}$, &c., which may be written

$a^{-1}-a^{-2}b+a^{-3}b^2-a^{-4}b^3+$, &c., the same as before found; and it is obvious, from inspecting the operation of division, that the series will never terminate.

Ex. 2. It is required to convert $\frac{1}{(a+b)^2}$ or $(a+b)^{-2}$ into an infinite series.

Ans. $\frac{1}{a^2}-\frac{2b}{a^3}+\frac{3b^2}{a^4}-\frac{4b^3}{a^5}+\frac{5b^4}{a^6}-$, &c.;

or, $a^{-2}-2a^{-3}b+3a^{-4}b^2-4a^{-5}b^3+5a^{-6}b^4$, &c.

Here the coefficients increase regularly by 1, and the signs are alternately positive and negative. We might have obtained the same result by division, as in the former example.

Ex. 3. Expand into a series $\frac{1}{a-b}$ or $(a-b)^{-1}$.

Here the coefficients furnished by the Rule are

$+1, \ -1, \ +1, \ -1$, &c.

But the factor b being negative, all its odd powers are negative. Hence the second term contains two negative factors, so that its resulting sign is $+$. The same remark applies to the fourth and sixth terms, &c., making the terms of the series all positive.

Ex. 4. Expand into a series $\frac{1}{(a-b)^2}$ or $(a-b)^{-2}$.

Ex. 5. Expand into a series $(a+b)^{-3}$.

Ans. $a^{-3}-3a^{-4}b+6a^{-5}b^2-10a^{-6}b^3+15a^{-7}b^4-$, &c.

Ex. 6. Expand into a series $(a-b)^{-4}$.

Ans. $a^{-4}+4a^{-5}b+10a^{-6}b^2+20a^{-7}b^3+35a^{-8}b^4+$, &c.

(272.) We have now considered the powers of a binomial when the exponent is an integer, either positive or negative. It remains to consider the case when the exponent is a *fraction.*

EXAMPLES.

Ex. 1. Expand $\sqrt{a+b}$ or $(a+b)^{\frac{1}{2}}$ into an infinite series.

The terms without the coefficients are

$$a^{\frac{1}{2}},\ a^{-\frac{1}{2}}b,\ a^{-\frac{3}{2}}b^2,\ a^{-\frac{5}{2}}b^3,\ a^{-\frac{7}{2}}b^4,\ \&c.$$

The exponents of a decrease by unity, while those of b increase by unity.

The coefficient of the first term is 1.

" second " $+\frac{1}{2}$.

" third " $\frac{\frac{1}{2}\times-\frac{1}{2}}{2}=-\frac{1}{2.4}$.

" fourth " $\frac{-\frac{1}{2.4}\times-\frac{3}{2}}{3}=+\frac{1.3}{2.4.6}$.

" fifth " $\frac{\frac{1.3}{2.4.6}\times-\frac{5}{2}}{4}=-\frac{1.3.5}{2.4.6.8}$.

The series, therefore, is

$$(a+b)^{\frac{1}{2}}=a^{\frac{1}{2}}+\frac{1}{2}a^{-\frac{1}{2}}b-\frac{1}{2.4}a^{-\frac{3}{2}}b^2+\frac{1.3}{2.4.6}a^{-\frac{5}{2}}b^3-\frac{1.3.5}{2.4.6.8}a^{-\frac{7}{2}}b^4+,\ \&c$$

The factors which form the coefficients are kept distinct, in order to show more clearly the *law* of the series. The numerators of the coefficients contain the series of odd numbers, 1, 3, 5, 7, &c., while the denominators contain the even numbers, 2, 4, 6, 8, 10, &c.

The above series expresses the square root of $a+b$. We shall obtain the same result if we extract the square root by the usual method. See *Art.* 299.

Ex. 2. It is required to convert $(a^2+x)^{\frac{1}{2}}$ into an infinite series.

$$Ans.\ a+\frac{a^{-1}x}{2}-\frac{a^{-3}x^2}{2.4}+\frac{3a^{-5}x^3}{2.4.6}-\frac{3.5.a^{-7}x^4}{2.4.6.8}+\frac{3.5.7a^{-9}x^5}{2.4.6.8.10}-, \&c.,$$

$$\text{or,}\quad a+\frac{x}{2a}-\frac{x^2}{2.4a^3}+\frac{3x^3}{2.4.6a^5}-\frac{3.5x^4}{2.4.6.8a^7}+\frac{3.5.7x^5}{2.4.6.8.10a^9}-, \&c.,$$

where the law of the series is evident.

Ex. 3. It is required to convert $(a-x)^{\frac{1}{2}}$ into an infinite series.

Ex. 4. It is required to convert $(a+b)^{\frac{1}{3}}$ into an infinite series.

$$Ans.\ a^{\frac{1}{3}}\left\{1+\frac{b}{3a}-\frac{2b^2}{3.6a^2}+\frac{2.5b^3}{3.6.9a^3}-\frac{2.5.8b^4}{3.6.9.12a^4}+, \&c.\right\}.$$

Ex. 5. Expand $(a-b)^{\frac{1}{4}}$ into an infinite series.

$$Ans.\ a^{\frac{1}{4}}\left\{1-\frac{b}{4a}-\frac{3b^2}{4.8a^2}-\frac{3.7b^3}{4.8.12a^3}-\frac{3.7.11b^4}{4.8.12.16a^4}-, \&c.\right\}.$$

Ex. 6. Expand $(a+x)^{\frac{2}{3}}$ into an infinite series.

Ex. 7. Expand $(1-x)^{\frac{1}{5}}$ into an infinite series.

$$Ans.\ 1-\frac{1}{5}x-\frac{1.4}{5.10}x^2-\frac{1.4.9}{5.10.15}x^3-\frac{1.4.9.14}{5.10.15.20}x^4-, \&c.$$

Ex. 8. Expand $(a^3-b^3)^{\frac{1}{3}}$ into an infinite series.

$$Ans.\ a\left(1-\frac{b^3}{3a^3}-\frac{2b^6}{3.6a^6}-\frac{2.5b^9}{3.6.9a^9}-, \&c.\right)$$

(273.) The binomial theorem is also applicable to cases in which the value of the exponent n is a *negative fraction.*

EXAMPLES.

Ex. 1. Expand into a series $\frac{1}{(a+b)^{\frac{1}{2}}}$ or $(a+b)^{-\frac{1}{2}}$.

The terms without the coefficients are

$$a^{-\frac{1}{2}},\ a^{-\frac{3}{2}}b,\ a^{-\frac{5}{2}}b^2,\ a^{-\frac{7}{2}}b^3,\ a^{-\frac{9}{2}}b^4,\ a^{-\frac{11}{2}}b^5, \&c.$$

The coefficient of the first term is 1.

" second " $-\frac{1}{2}$.

The coefficient of the third term is $\frac{-\frac{1}{2}\times-\frac{3}{2}}{2}=+\frac{1.3}{2.4}$.

" fourth " $\frac{\frac{1.3}{2.4}\times-\frac{5}{2}}{3}=-\frac{1.3.5}{2.4.6}$.

" fifth " $\frac{-\frac{1.3.5}{2.4.6}\times-\frac{7}{2}}{4}=+\frac{1.3.5.7}{2.4.6.8}$.

Hence we obtain

$$(a+b)^{-\frac{1}{2}}=a^{-\frac{1}{2}}-\frac{1}{2}a^{-\frac{3}{2}}b+\frac{1.3}{2.4}a^{-\frac{5}{2}}b^2-\frac{1.3.5}{2.4.6}a^{-\frac{7}{2}}b^3+\frac{1.3.5.7}{2.4.6.8}a^{-\frac{9}{2}}b^4-$$
&c.

Ex. 2. Expand into an infinite series $(a+x)^{-\frac{1}{3}}$.

Ans. $a^{-\frac{1}{3}}-\frac{1}{3}a^{-\frac{4}{3}}x+\frac{1.4}{3.6}a^{-\frac{7}{3}}x^2-\frac{1.4.7}{3.6.9}a^{-\frac{10}{3}}x^3+\frac{1.4.7.10}{3.6.9.12}a^{-\frac{13}{3}}x^4-$
&c.

Ex. 3. Expand $(1+x)^{-\frac{1}{5}}$ into an infinite series.

Ans. $1-\frac{x}{5}+\frac{6x^2}{5.10}-\frac{6.11x^3}{5.10.15}+\frac{6.11.16x^4}{5.10.15.20}-$, &c.

Ex. 4. Expand $(a^2-x)^{-\frac{1}{2}}$ into an infinite series.

Ans. $\frac{1}{a}+\frac{x}{2a^3}+\frac{1.3x^2}{2.4a^5}+\frac{1.3.5x^3}{2.4.6a^7}+\frac{1.3.5.7x^4}{2.4.6.8a^9}+$, &c.

Ex. 5. Expand $\frac{m}{\sqrt{b^2+c^4}}$ into an infinite series.

Ans. $\frac{m}{b}\left(1-\frac{c^4}{2b^2}+\frac{1.3c^8}{2.4b^4}-\frac{1.3.5c^{12}}{2.4.6b^6}+\frac{1.3.5.7c^{16}}{2.4.6.8b^8}-, \&c.\right)$.

(274.) The binomial theorem may be employed to determine the roots of surd numbers.

EXAMPLES.

Ex. 1. It is required to find the square root of 2.

The development of $(a+b)^{\frac{1}{2}}$ has been given in *Ex.* 1, page 224. If we make $a=1$ and $b=1$, then $(a+b)^{\frac{1}{2}}$ becomes $(1+1)^{\frac{1}{2}}$ or $\sqrt{2}$; and the terms of the development become

$$1+\frac{1}{2}-\frac{1}{2.4}+\frac{1.3}{2.4.6}-\frac{1.3.5}{2.4.6.8}+\frac{1.3.5.7}{2.4.6.8.10}-, \&c.,$$

which therefore expresses the square root of 2. The sum of

this series is 1.41421. As, however, the series converges very slowly, it would require a large number of terms to give the root with tolerable accuracy. The following example affords a better illustration of the *utility* of the method.

Ex. 2. Required the square root of 101.

$$101=100\left(1+\frac{1}{100}\right). \quad \text{Therefore } \sqrt{101}=10\left(1+\frac{1}{100}\right)^{\frac{1}{2}}.$$

Put $a=1$ and $b=\frac{1}{100}$ in the development of $(a+b)^{\frac{1}{2}}$ on page 224, and we shall have

$$\sqrt{101}=10\left(1+\frac{1}{2.100}-\frac{1}{2.4.100^2}+\frac{1.3}{2.4.6.100^3}-\frac{1.3.5}{2.4.6.8.100^4}+, \&c.\right)$$

This series converges so rapidly that the first two terms give a result correct to three decimal places, and five terms give a result correct to ten decimal places.

Thus, the value of the first term is		1.00000000000
"	second "	+ .00500000000
"	third "	− .00001250000
"	fourth "	+ .00000006250
"	fifth "	− .00000000039
Their sum is		1.00498756211.

And multiplying by 10, we have

$$\sqrt{101}=10.0498756211.$$

Ex. 3. It is required to convert $\sqrt[3]{9}$, or its equal $(8+1)^{\frac{1}{3}}$, into an infinite series, and find its value.

Ans.

$$2+\frac{1}{3.2^2}-\frac{1}{3.6.2^4}+\frac{5}{3.6.9.2^7}-\frac{5.8}{3.6.9.12.2^{10}}+\frac{5.8.11}{3.6.9.12.15.2^{13}}-, \&c.,$$
$$=2.08008.$$

Ex. 4. It is required to extract the cube root of 31.

$$\sqrt[3]{31}=\sqrt[3]{27+4}=\sqrt[3]{27}\left(1+\frac{4}{27}\right)^{\frac{1}{3}},$$

$$=3\left\{1+\frac{4}{3.27}-\frac{2.4^2}{3.6.27^2}+\frac{2.5.4^3}{3.6.9.27^3}-\frac{2.5.8.4^4}{3.6.9.12.27^4}+, \&c.\right\},$$
$$=3.14138.$$

SECTION XVII.

EVOLUTION OF POLYNOMIALS.

(275.) *Method of extracting the square root of a polynomial.*

In order to discover a rule for extracting the square root, let us consider the square of $a+b$, which is $a^2+2ab+b^2$. If we write the terms of the square in such a manner that the powers of one of the letters, as a, may go on continually decreasing, the first term will be the square of the first term of the root; and since in the present case the first term of the square is a^2, the first term of the root must be a.

Having found the first term of the root, we must consider the rest of the square, namely, $2ab+b^2$, to see how we can derive from it the second term of the root. Now this remainder, $2ab+b^2$, may be put under the form $(2a+b)b$; whence it appears that we shall find the second term of the root if we divide the remainder by $2a+b$. The first part of this divisor, $2a$, is double of the first term already determined; the second part, b, is yet unknown, and it is necessary at present to leave its place empty. Nevertheless, we may commence the division, employing only the term $2a$; but as soon as the quotient is found, which, in the present case, is b, we must put it in the vacant place, and thus render the divisor complete.

The whole process, therefore, may be represented as follows:

$$\begin{array}{r|l} a^2+2ab+b^2 & \underline{a+b} = \text{the root.} \\ a^2 & \\ \hline 2ab+b^2 & 2a+b = \text{the divisor.} \\ 2ab+b^2 & \end{array}$$

Hence we derive the following

RULE FOR EXTRACTING THE SQUARE ROOT OF A POLYNOMIAL.

Arrange the terms according to the powers of some one letter; take the square root of the first term for the first term of the required root, and subtract its square from the given polynomial.

Divide the first term of the remainder by double the root already found, and annex the result both to the root and the divisor. Multiply the divisor thus increased by the last term of the root, and subtract the product from the last remainder. Proceed in the same manner to find the additional terms of the root.

Ex. 1. Required the square root of $a^4-2a^3x+3a^2x^2-2ax^3+x^4$.

$a^4-2a^3x+3a^2x^2-2ax^3+x^4 \mid a^2-ax+x^2$ = the root.
a^4
$\quad -2a^3x+3a^2x^2 \mid 2a^2-ax$ = the first divisor.
$\quad -2a^3x+\ \ a^2x^2 \mid$
$\qquad\qquad 2a^2x^2-2ax^3+x^4 \mid 2a^2-2ax+x^2$ = the second divisor.
$\qquad\qquad 2a^2x^2-2ax^3+x^4 \mid$

For verification, multiply the root a^2-ax+x^2 by itself, and we shall obtain the original polynomial.

Ex. 2. Required the square root of $a^2+2ab+2ac+b^2+2bc+c^2$.

Ex. 3. Required the square root of $10x^4-10x^3-12x^5+5x^2+9x^6-2x+1$.

Ex. 4. Required the square root of $8ax^3+4a^2x^2+4x^4+16b^2x^2+16b^4+16ab^2x$.

Ans. $2x^2+2ax+4b^2$.

Ex. 5. Extract the square root of $15a^4b^2+a^6-6a^5b-20a^3b^3+b^6+15a^2b^4-6ab^5$.

Ex. 6. Extract the square root of $8ab^3+a^4-4a^3b+4b^4$.

(276.) *Method of extracting the square root of numbers.*

The preceding rule is applicable to the extraction of the square root of numbers. For every number may be regarded as an Algebraic polynomial, or as composed of a certain number of units, tens, hundreds, &c. Thus,

529 is equivalent to 500+20+9.
Also, 841 " 800+40+1.

If, then, 841 is the square of a number composed of tens and units, it must contain *the square of the tens, plus twice the product of the tens by the units, plus the square of the units.* But these three terms are blended together in 841, and hence the peculiar difficulty in determining its root. The following principles will, however, enable us to separate these terms, and thus detect the root.

(277.) I. *For every two figures of the square there will be one figure in the root, and also one for any odd figure.*

Thus, the square of 1 is 1,
" 10 is 100,
" 100 is 10000,
" 1000 is 1000000,
&c., &c.

The smallest number consisting of two figures is 10, and its square is the smallest number of three figures. The smallest number of three figures is 100, and its square is the smallest number of five figures, and so on. Therefore, the square root of every number composed of one or two figures will contain *one* figure; the square root of every number composed of three or four figures will contain *two* figures; of a number from five to six figures will contain *three* figures; and from $2n-1$ to $2n$ figures must contain n figures.

Hence, if we divide the number into periods of two figures, proceeding from right to left, the number of figures in the root will be equal to the number of periods.

(278.) II. *The first figure of the root will be the square root of the greatest square number contained in the first period on the left.*

For the square of tens can give no figure in the first right hand period; the square of hundreds can give no figure in the first two periods on the right; and the square of the highest figure in the root can give no figure except in the first period on the left.

Ex. 1. Suppose we wish to find the square root of 529.

The square of 23 or $20+3$ is $20^2+2.20.3+3^2$,

or $$400+120+9.$$

Here the three classes of terms are exhibited distinct from each other, and we might extract the root by the rule of *Art.* 275. But observe that in the number 529, since the square of the tens can not give a figure in the place of units or tens, it must be contained in the first period 5. Now this period contains not only the square of the tens, but also a part of the product of the tens by the units. The greatest square contained in 5 is 4, whose root is 2; hence 2 must be the number of tens, whose square is 400; and if we subtract this from 529, the remainder 129 contains twice the product of the tens by the units, plus the square of the units. If, then, we divide this remainder by twice the tens, we shall obtain the units, or possibly a number somewhat too large. This quotient figure can never be too small, but it may be too large, because the remainder 129, besides twice the product of the tens by the units, contains the square of the units. We therefore complete the divisor by annexing the quotient 3 to the right of the 4, and then multiplying by 3, we evidently obtain the double product of the tens by the units, plus the square of the units. The entire operation may then be represented as follows:

```
     5·29|23=the root
     4
  43|129
     129.
```

(279.) Hence, for the extraction of the square root of numbers we derive the following

RULE.

1. *Separate the given number into periods of two figures each, beginning at the right hand.*

2. *Find the greatest square contained in the left-hand period; its root is the first figure of the required root. Subtract the square from the first period, and to the remainder bring down the second period for a dividend.*

3. *Double the root already found for a divisor, and find how many times it is contained in the dividend, exclusive of its right-hand figure; annex the result both to the root and the divisor.*

4. *Multiply the divisor thus increased by the last figure of the root; subtract the product from the dividend, and to the remainder bring down the next period for a new dividend.*

5. *Double the whole root now found for a new divisor, and continue the operation, as before, until the periods are all brought down.*

Ex. 2. Find the square root of 186624.

The operation is as follows:

```
     18·66·24|432
     16
 83|  2 66
      2 49
862|  17 24
      17 24.
```

Ex. 3. Find the square root of 21086464.

Ex. 4. Find the square root of 88078225.

(280.) We have seen that the root of a fraction is equal to the root of its numerator divided by the root of its denominator. Hence the square root of $\frac{529}{100}$, or 5.29, is $\frac{23}{10}$, or 2.3.

The square root of $\frac{186624}{10000}$, or 18.6624, is $\frac{432}{100}$, or 4.32.

That is, the square root of a decimal fraction may be found in the same manner as a whole number, if we divide it into periods commencing with the decimal point.

Ex. 5. Find the square root of 58.614336.

Ex. 6. Find the square root of 9.878449.

Hence, also, if the square root of a number can not be found exactly, we may, by annexing ciphers, obtain the root approximately in decimal fractions.

Ex. 7. Find the square root of 2.

Ans. 1.4142136 nearly.

Ex. 8. Find the square root of 3.

Ex. 9. Find the square root of 10.

The most expeditious method of extracting roots is usuallv by means of logarithms. See page 308.

(281.) *Method of extracting the cube root of a polynomial.*

We already know that the cube of $a+b$ is $a^3+3a^2b+3ab^2+b^3$

If, then, the *cube* were given, and we were required to find its *root*, it might be done by the following method:

When the terms are arranged according to the powers of one letter, we at once know, from the first term a^3, that a must be one term of the root. If, then, we subtract its cube from the proposed polynomial, we obtain the remainder $3a^2b+3ab^2+b^3$, which must furnish the second term of the root.

Now this remainder may be put under the form

$$(3a^2+3ab+b^2)\times b;$$

whence it appears that we shall find the second term of the root, if we divide the remainder by $3a^2+3ab+b^2$. But as this second term is supposed to be unknown, the divisor can not be completed. Nevertheless, we know the first term $3a^2$, that is, thrice the square of the first term already found, and by means of this we can find the other part b, and then complete the divisor before we perform the division. For this purpose, we must add to $3a^2$ thrice the product of the two terms, or $3ab$, and the square of the second term of the root, or b^2. Hence we derive the following

RULE FOR EXTRACTING THE CUBE ROOT OF A POLYNOMIAL.

(282.) *Arrange the terms according to the powers of some one letter, take the cube root of the first term, and subtract the cube from the given polynomial.*

Divide the first term of the remainder by three times the square of the root already found, the quotient will be the second term of the root.

Complete the divisor by adding to it three times the product of the two terms of the root, and the square of the second term.

Multiply the divisor thus increased by the last term of the root, and subtract the product from the last remainder. Proceed in the same manner to find the additional terms of the root.

Ex. 1. Extract the cube root of $a^3+12a^2+48a+64$.

$$\begin{array}{l|l} a^3+12a^2+48a+64 & a+4 = \text{the root.} \\ a^3 & \\ \hline \quad 12a^2+48a+64 & 3a^2+12a+16 = \text{the divisor.} \\ \quad 12a^2+48a+64 & \\ \hline \quad .. \quad .. \quad .. & \end{array}$$

Having found the first term of the root a, and subtracted its cube, we divide the first term of the remainder, $12a^2$, by three times the square of a, that is, $3a^2$, and we obtain 4 for the second term of the root. We then complete the divisor by adding to it three times the product of the two terms of the root, which is $12a$, together with the square of the last term, 4, which is 16. Multiplying, then, the complete divisor by 4, and subtracting the product from the last remainder, nothing is left. Hence the required cube root is $a+4$.

This result may be easily verified by multiplication.

Ex. 2. Extract the cube root of $a^6-6a^5+15a^4-20a^3+15a^2-6a+1$.

$$
\begin{array}{l|l}
a^6-6a^5+15a^4-20a^3+15a^2-6a+1 & a^2-2a+1 = \text{the root.} \\
a^6 & \\
\hline
\quad -6a^5+15a^4-20a^3 & 3a^4-6a^3+4a^2 = \text{the first divisor.} \\
\quad -6a^5+12a^4-\ \ 8a^3 & \\
\hline
\qquad\qquad 3a^4-12a^3+15a^2-6a+1 & 3a^4-12a^3+15a^2-6a+1 = \\
\qquad\qquad 3a^4-12a^3+15a^2-6a+1 & \qquad\qquad \text{the second divisor.} \\
\hline
\qquad\qquad \cdot\cdot \quad \cdot\cdot \quad \cdot\cdot \quad \cdot\cdot \quad \cdot\cdot &
\end{array}
$$

We may dispense with forming the complete divisor according to the rule, if, each time that we find a new term of the root, we raise the entire root to the third power, and subtract the cube from the given polynomial.

Ex. 3. Required the cube root of $6x^5-40x^3+x^6+96x-64$.

Ex. 4. Required the cube root of $18x^4+36x^2+24x+8+32x^3+x^6+6x^5$.

Ex. 5. Required the cube root of $3b^5+b^6-5b^3-1+3b$.

(283.) *Method of extracting the cube root of numbers.*

The preceding rule is applicable to the extraction of the cube root of numbers; but a difficulty in applying it arises from the fact that the terms of the power are all blended together in the given number. They may, however, be separated by attending to the following principles:

I. *For every three figures of the cube there will be one figure in the root, and also one for any additional figure or figures.*

Thus, the cube of 1 is 1,
" 10 is 1000,
" 100 is 1000000,
" 1000 is 1000000000,
&c., &c.

Hence we see that the cube root of a number consisting of from 1 to 3 figures will contain *one* figure; of a number from 4 to 6 figures will contain *two* figures; from 7 to 9 figures will contain *three;* and from $3n-2$ to $3n$ figures must contain n

figures. Hence, if we divide the number into periods of three figures, proceeding from right to left, the number of figures in the root will be equal to the number of periods.

II. *The first figure of the root will be the cube root of the greatest cube number contained in the first period on the left.*

Ex. 1. Suppose we wish to find the cube root of 12167.

The cube of 23 or $20+3$ is $20^3+3.20^2.3+3.20.3^2+3^3$,

or $$8000+3600+540+27.$$

Here the four classes of terms are exhibited distinct from each other, and the rule of *Art.* 282 might be easily applied. But observe that in the number 12167, since the cube of the tens can not give a figure in the first three places, it must be contained in the first period 12. The greatest cube contained in this is 8, the root of which is 2. Hence 2 must be the number of tens whose cube is 8000; and the remainder 4167 contains *three times the product of the square of the tens by the units, plus three times the product of the tens by the square of the units, plus the cube of the units.*

If, then, we divide this remainder by three times the square of the tens, we shall obtain the units, or possibly a number too large, because the divisor is too small. We therefore complete the divisor by adding to it three times the product of the tens by the units, plus the square of the units. The entire operation is then as follows:

		12·167 | 23 = the root.
		8
$20^2\times 3$	$=1200$	4 167
$20\times 3\times 3$	$=180$	
3^2	$=9$	4 167
complete divisor	$=1389$	

(284.) Hence, for the extraction of the cube root of numbers, we derive the following

RULE.

1. *Separate the given number into periods of three figures each, beginning at the right hand.*

2. *Find the greatest cube contained in the left-hand period; its root is the first figure of the required root. Subtract the cube*

from the first period, and to the remainder bring down the second period for a dividend.

3. *Take three hundred times the square of the root already found for a trial divisor; find how many times it is contained in the dividend, and place the quotient for a second figure of the root.*

4. *Complete the divisor by adding to it thirty times the product of the two figures of the root, and the square of the second figure.*

5. *Multiply the divisor thus increased by the last figure of the root; subtract the product from the dividend, and to the remainder bring down the next period for a new dividend.*

6. *Take three hundred times the square of the whole root now found for a new trial divisor, and continue the operation as before until all the periods are brought down.*

It will be observed that three times the square of the tens, when their local value is regarded, is the same as three hundred times the square of this digit, not regarding its local value.

Ex. 2. Find the cube root of 20796875.

Ex. 3. Find the cube root of 2509911279.

Ex. 4. Find the cube root of 895562584119.

The same method is applicable to the extraction of the cube root of fractions, and also of imperfect powers.

Ex. 5. Find the cube root of 604.422796375.

Ex. 6. Find the cube root of 4.

Ex. 7. Find the cube root of 11.

(285.) *Method of extracting any root of a polynomial.*

We already know that the nth power of $a+b$ is $a^n+na^{n-1}b+$ other terms. The first term of the root is, therefore, the nth root of the first term of the polynomial. Also, the second term of the root may be found by dividing the second term of the polynomial by na^{n-1}; that is, the first term of the root raised to the next inferior power, and multiplied by the exponent of the given power. Hence we deduce the following

RULE FOR EXTRACTING ANY ROOT OF A POLYNOMIAL.

Arrange the terms according to the powers of one of the letters, and take the n*th root of the first term for the first term of the required root.*

Subtract its power from the given polynomial, and divide the first term of the remainder by n *times the* (n—1) *power of this root; the quotient will be the second term of the root.*

Subtract the n*th power of the terms already found from the given quantity, and, using the same divisor, proceed in like manner to find the remaining terms of the root.*

Ex. 1. Required the fourth root of $16a^4-96a^3x+216a^2x^2-216ax^3+81x^4$.

$16a^4-96a^3x+216a^2x^2-216ax^3+81x^4 \,|\, \underline{2a-3x}$ = the root.
$16a^4$
$-96a^3x \,|\, 32a^3$ = the divisor.
$16a^4-96a^3x+216a^2x^2-216ax^3+81x^4$.

Here we take the fourth root of $16a^4$, which is $2a$, for the first term of the required root; subtract its fourth power, and bring down the first term of the remainder $-96a^3x$. For a divisor, we raise the first term of the root to the third power, and multiply it by 4, making $32a^3$. Dividing, we obtain $-3x$ for the second term of the root. The quantity $2a-3x$ being raised to the fourth power, is found to be equal to the proposed polynomial.

Ex. 2. Required the fifth root of $80x^3+32x^5-80x^4-40x^2+10x-1$.

Ans. $2x-1$.

Ex. 3. Required the fourth root of $336x^6+81x^8-216x^7-56x^4+16-224x^3+64x$.

Ans. $3x^2-2x-2$.

(286.) *Method of extracting any root of numbers.*

It is easy to apply the preceding Rule to the extraction of any root of numbers. For a reason similar to that given for the square and cube roots, we must first divide the number into periods of n figures each. Then the first figure of the root will be the nth root of the greatest nth power contained in the first period on the left. If we subtract its power from the given number, and divide the remainder by n times the $(n—1)$ power of the first figure, regarding its local value, the quotient will be the second figure of the root, or, possibly, something too large. The result may be verified by raising the whole root now found to the nth power; and if there are other figures, they may be found in the same manner.

Ex. 1. Find the fifth root of 33554432.

$$\begin{array}{rl} & 335\cdot54432 \,|\, 32 \\ & \underline{243} \\ 5.3^4=405 & | \; 925 \\ 32^5= & 335\;54432. \end{array}$$

Ex. 2. Find the fifth root of 4984209207.

Ex. 3. Find the fifth root of 10.

(287.) When the index of the root to be extracted is a multiple of two or more numbers, we may obtain the root required by the successive extraction of simpler roots, *Art.* 159.

For example, we may obtain the *fourth* root by extracting the square root *twice* successively; for the square root of a^4 is a^2, and the square root of a^2 is a.

The *eighth* root may be obtained by extracting the square root *three times* successively; for the square root of a^8 is a^4, that of a^4 is a^2, and that of a^2 is a.

In the same manner, the *sixteenth* root may be obtained by extracting the square root *four* times successively, and so on.

The *sixth* root may be found by extracting the square root, and afterward the cube root; for the square root of a^6 is a^3, and the cube root of a^3 is a. We may also take, first, the cube root, which gives a^2, and afterward the square root, which gives a, as before. It is, however, best to extract the roots of the lowest degree first, because the operation is less laborious.

In general, *the* m*n*th *root of a number is equal to the* n*th root of the* m*th root of this number.* That is,

$$\sqrt[mn]{a}=\sqrt[n]{\sqrt[m]{a}}.$$

For, raising each member of this equation to the nth power we have

$$\sqrt[m]{a}=\sqrt[m]{a}.$$

Ex. 1. Find the fourth root of $6a^2b^2+a^4-4a^3b-4ab^3+b^4$.

Ex. 2. Find the sixth root of $6a^5b+15a^4b^2+a^6+20a^3b^3+15a^2b^4+b^6+6ab^5$.

Ex. 3. Find the eighth root of $1024x^7y+1792x^6y^2+256x^8+1120x^4y^4+1792x^5y^3+448x^3y^5+y^8+112x^2y^6+16xy^7$.

EXTRACTION OF THE SQUARE ROOT OF A QUANTITY OF THE FORM $a\pm\sqrt{b}$.

(288.) Binomials of this class require particular attention, because they frequently occur in the solution of equations of the fourth degree, such as are treated of in *Art.* 184. Thus the equation

$$x^4=14x^2-1,$$

gives us $$x^2=7\pm4\sqrt{3}.$$

Hence, in order to find the value of x, we must extract the square root of the binomial $7\pm4\sqrt{3}$.

In order to show that the square root of such an expression may sometimes be extracted, take the binomial

$$2\pm\sqrt{3},$$

and find its square.

$$(2\pm\sqrt{3})^2=4\pm4\sqrt{3}+3=7\pm4\sqrt{3}.$$

Therefore, the square root of $7\pm4\sqrt{3}$ is $2\pm\sqrt{3}$.

The square root of an expression of the form $a\pm\sqrt{b}$ may, therefore, sometimes be extracted, and it is required to determine a general method for this purpose whenever it is practicable.

THEOREM I.

(289.) *The sum or difference of two surds can not be equal to a rational quantity.*

For, if possible, let $\sqrt{a}\pm\sqrt{b}=c$, where c denotes a rational quantity, and $\sqrt{a}$, $\sqrt{b}$ denote surd quantities.

By transposing $\sqrt{b}$ and squaring both sides, we obtain $a=c^2\mp2c\sqrt{b}+b$; whence, by transposition and division, we have

$$\pm\sqrt{b}=\frac{b+c^2-a}{2c}.$$

The second member of the equation contains only rational quantities, while $\sqrt{b}$ was supposed to be *irrational;* that is, we find an irrational quantity equal to a rational one, which is absurd. Hence the sum or difference of two surds can not be equal to a rational quantity.

THEOREM II.

In every equation of the form

$$x \pm \sqrt{y} = a \pm \sqrt{b},$$

the rational parts on the opposite sides are equal to each other, and also the irrational parts.

For if x is not equal to a, let it be equal to $a \pm z$

Then $$a \pm z \pm \sqrt{y} = a \pm \sqrt{b};$$

or $$z = \sqrt{b} - \sqrt{y};$$

that is, a rational quantity is equal to the difference of two surds, which, by the last Theorem, is impossible. Therefore, $x=a$, and, consequently, $\sqrt{y}=\sqrt{b}$.

THEOREM III.

If $\sqrt{a+\sqrt{b}}$ *is equal to* $x+\sqrt{y}$,

then will $\sqrt{a-\sqrt{b}}$ *be equal to* $x-\sqrt{y}$.

For, by involution, $a+\sqrt{b}=x^2+2x\sqrt{y}+y$.

But, by the last Theorem, $a=x^2+y$,

and $\sqrt{b}=2x\sqrt{y}$.

Subtracting, we obtain $a-\sqrt{b}=x^2-2x\sqrt{y}+y$.

Therefore, by evolution, $\sqrt{a-\sqrt{b}}=x-\sqrt{y}$.

(290.) *To find an expression for the square root of* $a \pm \sqrt{b}$.

Let us assume $\sqrt{a+\sqrt{b}}=p+q$ (1),

where p and q may be both radicals, or one rational and the other a radical, but p^2 and q^2 are required to be *rational.*

Then, by the last Theorem,

$$\sqrt{a-\sqrt{b}}=p-q \quad (2).$$

Multiplying these equations together, we obtain

$$\sqrt{a^2-b}=p^2-q^2 \quad (3), \text{ a rational quantity.}$$

Hence we see that, in order that $\sqrt{a+\sqrt{b}}$ may be expressed by the sum of two radicals, or one rational term and the other a radical, *the expression* a^2-b *must be a perfect square.*

Let, then, a^2-b be a perfect square, and put $\sqrt{a^2-b}=c$; equation (3) will thus become

$$p^2-q^2=c.$$

Squaring equations (1) and (2), we obtain

$$p^2+q^2+2pq=a+\sqrt{b},$$
$$p^2+q^2-2pq=a-\sqrt{b}.$$

Adding these two equations, we obtain

$$p^2+q^2=a.$$

But we have already found

$$p^2-q^2=c.$$

Hence $$2p^2=a+c,$$
and $$2q^2=a-c.$$

From which we obtain

$$p=\pm\sqrt{\frac{a+c}{2}},$$

and $$q=\pm\sqrt{\frac{a-c}{2}}.$$

Therefore,

$$\sqrt{a+\sqrt{b}},\text{ or }p+q=\pm\left(\sqrt{\frac{a+c}{2}}+\sqrt{\frac{a-c}{2}}\right)$$

$$\sqrt{a-\sqrt{b}},\text{ or }p-q=\pm\left(\sqrt{\frac{a+c}{2}}-\sqrt{\frac{a-c}{2}}\right).$$

(291.) Hence, to extract the square root of a binomial of the form $a\pm\sqrt{b}$, we have the following

RULE.

From the square of the rational part (a^2), *take the square of the irrational part* (b); *extract the square root of the remainder and, calling that root* c, *the required root will be*

$$\sqrt{\frac{a+c}{2}}\pm\sqrt{\frac{a-c}{2}}.$$

Ex. 1. Required the square root of $4+2\sqrt{3}$.

Here $a=4$, and $\sqrt{b}=2\sqrt{3}$; therefore, $a^2-b=c^2=16-12=4$; or $c=2$. Hence, by the above formula, the required root will be

$$\sqrt{\frac{4+2}{2}}+\sqrt{\frac{4-2}{2}}=\sqrt{3}+1.$$

Verification.

The square of $\sqrt{3}+1$ is $3+2\sqrt{3}+1=4+2\sqrt{3}$.

Ex. 2. Required the square root of $11+6\sqrt{2}$.

Here $a=11$, and $\sqrt{b}=6\sqrt{2}$; therefore, $b=36\times2=72$; and $a^2-b=49=c^2$. Hence $c=7$, and we find the square root of $11+6\sqrt{2}$ is $\sqrt{9}+\sqrt{2}$, or $3+\sqrt{2}$. *Ans.*

Ex. 3. Required the square root of $11-2\sqrt{30}$.

Ans. $\sqrt{6}-\sqrt{5}$

Ex. 4. Required the square root of $2+\sqrt{3}$.

Ans. $\sqrt{\tfrac{3}{2}}+\sqrt{\tfrac{1}{2}}$.

(292.) This method is applicable even when the binomial contains imaginary quantities.

Ex. 5. Required the square root of $1+4\sqrt{-3}$.

Here $a=1$, and $\sqrt{b}=4\sqrt{-3}$; hence $b=-48$, and $a^2-b=49$; therefore, $c=7$. The required square root is $\sqrt{4}+\sqrt{-3}=2+\sqrt{-3}$. *Ans.*

Ex. 6. Required the square root of $-\tfrac{1}{2}+\tfrac{1}{2}\sqrt{-3}$.

Ans. $\tfrac{1}{2}+\tfrac{1}{2}\sqrt{-3}$.

Ex. 7. Required the square root of $2\sqrt{-1}$.

Here we put $a=0$; hence $c=2$, and the required root is

$$1+\sqrt{-1},$$

which may be easily verified.

Ex. 8. Required the value of the expression

$$\sqrt{6+2\sqrt{5}}-\sqrt{6-2\sqrt{5}}.$$

Ex. 9. Required the value of the expression

$$\sqrt{4+3\sqrt{-20}}+\sqrt{4-3\sqrt{-20}}.$$

Ans. 6.

SECTION XVIII.

INFINITE SERIES.

(293.) An *infinite series* is an infinite number of terms, each of which is derived from the preceding term or terms according to some law.

As $$1+\frac{1}{2}+\frac{1}{4}+\frac{1}{8}+\frac{1}{16}+, \text{\&c.},$$

or $$1-\frac{1}{3}+\frac{1}{9}-\frac{1}{27}+\frac{1}{81}-, \text{\&c.}$$

These are examples of geometrical progressions, in the first of which the ratio is $\frac{1}{2}$, and in the second it is $-\frac{1}{3}$.

Infinite series may arise from the common operations of division, the extraction of roots, and other processes of calculation, as will be seen hereafter.

A converging series is one in which the sum of any number of its terms is finite, as in the examples just given.

A diverging series is one in which the sum of its terms is not finite; as,

$$1+2+3+4+5+6+7, \text{\&c.}$$

An ascending series is one in which the exponents of the unknown quantity continually increase; as,

$$ax+bx^2+cx^3+dx^4+ex^5+, \text{\&c.}$$

A descending series is one in which the exponents of the unknown quantity continually decrease; as,

$$ax^{-1}+bx^{-2}+cx^{-3}+dx^{-4}+ex^{-5}+, \text{\&c.}$$

PROBLEM I.

(294.) *Any series being given, to find its several orders of differences.*

RULE.

1. *Take the first term from the second, the second from the third, the third from the fourth, &c.; and the remainders will form a new series, called the* FIRST ORDER OF DIFFERENCES.

2. *Take the first term of this last series from the second, the second from the third, &c.; and the remainders will form a third series, called the* SECOND ORDER OF DIFFERENCES.

3. *Proceed in like manner for the third, fourth, &c., orders of differences, and so on till they terminate, or are carried as far as may be thought necessary.*

Ex. 1. Required the several orders of differences of the series of squares,

1	4	9	16	25	36	49, &c.	
3	5	7	9	11	13		first differences.
2	2	2	2	2			second differences.
0	0	0	0				third differences.

Ex. 2. Required the several orders of differences of the series of cubes,

1	8	27	64	125	216, &c.	
7	19	37	61	91		first differences.
12	18	24	30			second differences.
6	6	6				third differences.
0	0					fourth differences.

Ex. 3. Required the several orders of differences of the series of fourth powers,

1, 16, 81, 256, 625, 1296, &c.

Ex. 4. Required the several orders of differences of the series of fifth powers,

1, 32, 243, 1024, 3125, 7776, 16807, &c.

Ex. 5. Required the several orders of differences of the series of numbers,

1, 3, 6, 10, 15, 21, &c.

PROBLEM II.

(295.) *To find the* n*th term of the series*

$a, b, c, d, e,$ &c.

Take the proposed series, and subtract each term from the next succeeding one; we shall thus obtain for the *first order of differences,*

$$b-a,\ c-b,\ d-c,\ e-d,\ \&c.$$

Again, subtracting each term of this series from the next succeeding term, we find for the *second order of differences,*

$$c-2b+a,\ d-2c+b,\ e-2d+c,\ \&c.$$

Subtracting, again, each term of the preceding series from ts next succeeding term, we find the *third order of differences,*

$$d-3c+3b-a,\ e-3d+3c-b,\ \&c.$$

Subtracting again, we find for the *fourth order of differences,*

$$e-4d+6c-4b+a,\ \&c.$$

Let D', D'', D''', D'''', &c., represent the first terms of the several orders of differences.

Then,

$D' = b-a$; whence $b=a+D'$,

$D'' = c-2b+a$; " $c=a+2D'+D''$,

$D''' = d-3c+3b-a$; " $d=a+3D'+3D''+D'''$,

$D''''=e-4d+6c-4b+a$; " $e=a+4D'+6D''+4D'''+D''''$,

&c., &c.

The coefficients of the value of c, the *third* term of the proposed series, are 1, 2, 1, which are the coefficients of the *second* power of a binomial; the coefficients of the value of d, the *fourth* term, are 1, 3, 3, 1, which are the coefficients of the *third* power of a binomial, and so on. Hence we infer that the coefficients of the nth term of the series are the coefficients of the $(n-1)$ power of a binomial. Therefore, the nth term of the series will be

$$a+\frac{(n-1)}{1}D'+\frac{(n-1)(n-2)}{1.2}D''+\frac{(n-1)(n-2)(n-3)}{1.2.3}D'''+,\ \&c.$$

Ex. 1. Required the twelfth term of the series

2, 6, 12, 20, 30, &c.

The first order of differences is 4 6 8 10, &c.
The second order of differences is 2 2 2, &c.
The third order of differences is 0 0.

Here $D'=4$, $D''=2$, and $D'''=0$. Also, $a=2$, and $n=12$.

Hence $a+(n-1)D'+\frac{(n-1)(n-2)D''}{2}=2+11D'+55D''=$

$2+44+110=156=$ the twelfth term.

Ex. 2. Required the twentieth term of the series

1, 3, 6, 10, 15, 21, &c.

Here $a=1$, $D'=2$, $D''=1$, and $n=20$.

Therefore, the 20th term $=1+19D'+171D''=1+38+171$ $=210$, *Ans.*

Ex. 3. Required the thirteenth term of the series

1, 5, 14, 30, 55, 91, &c.

Ex. 4. Required the fifteenth term of the series

1, 4, 9, 16, 25, 36, &c.

Ans. 225.

Ex. 5. Required the twentieth term of the series

1, 8, 27, 64, 125, &c.

PROBLEM III.

(296.) *To find the sum of* n *terms of the series*

$a, b, c, d, e,$ &c.

Assume the series

$0, a, a+b, a+b+c, a+b+c+d,$ &c.

Subtracting each term from the next succeeding, we obtain

$a, b, c, d, e,$ &c.,

which is the series whose sum it is proposed to find. Hence the sum of n terms of the proposed series is the $(n+1)$th term of the assumed series; and the nth order of differences in the first series is the $(n+1)$th order in the other series. If, therefore, in the formula of the preceding Problem, we substitute

0 for a,
$n+1$ for n,
a for D',
D' for D'',
&c.,

we shall have

$$na+\frac{n(n-1)}{2}D'+\frac{n(n-1)(n-2)}{2.3}D''+\frac{n(n-1)(n-2)(n-3)}{2.3.4}D'''+$$

&c.,

which is the sum of n terms of the proposed series.

Ex. 1. Required the sum of n terms of the series

$$1, 2, 3, 4, 5, 6, \&c.$$

Here $a=1$, $D'=1$, $D''=0$.

Therefore, $na+\frac{n(n-1)D'}{2}=n+\frac{n^2-n}{2}=\frac{n^2+n}{2}=\frac{n(n+1)}{2}=$ the sum of n terms, the same as found in *Art.* 239.

Ex. 2. Required the sum of n terms of the series

$$1^2, 2^2, 3^2, 4^2, 5^2, \&c.$$

Here $a=1$, $D'=3$, $D''=2$.

Therefore the general formula reduces to

$$n+\frac{3n(n-1)}{2}+\frac{2n(n-1)(n-2)}{2.3},$$

$$=\frac{2n^3+3n^2+n}{6},$$

$$=\frac{n(n+1)(2n+1)}{6}, \text{ the sum required.}$$

Ex. 3. Required the sum of n terms of the series

$$1^3, 2^3, 3^3, 4^3, 5^3, 6^3, \&c.$$

Here $a=1$, $D'=7$, $D''=12$, $D'''=6$.

Ans. $\frac{n^2(n+1)^2}{4}$

Ex. 4. Required the sum of n terms of the series

$$1, 3, 6, 10, 15, \&c.$$

Ans. $\frac{n(n+1)(n+2)}{2.3}$

Ex. 5. Required the sum of n terms of the series

$$1, 4, 10, 20, 35, \&c.$$

Ans. $\frac{n(n+1)(n+2)(n+3)}{2.3.4}$.

PROBLEM IV.

(297.) *A series of equidistant terms,* a, b, c, d, e, *&c., being given, to find any intermediate term by interpolation.*

This is essentially the same as Problem II. For convenience, let us put x to represent the distance of the required term from the first term of the series a, in which case $x=n-1$, and we shall have

$$z=a+x\text{D}'+\frac{x(x-1)}{2}\text{D}''+\frac{x(x-1)\,(x-2)}{2.3}\text{D}'''+, \text{ \&c.}$$

Ex. 1. Given the square root of 94, equal to 9.69536;
" " 95, " 9.74679;
" " 96, " 9.79796,

to find the square root of 94.25.

Here the first differences are +.05143, +.05117,
and the second difference is −.00026;
that is, $\text{D}'=+.05143$, $\text{D}''=-.00026$.
But $z=a+\frac{1}{4}\text{D}'-\frac{3}{32}\text{D}''$.

Hence the square root of 94.25 is
9.69536+.01286+.00002,
or 9.70824. *Ans.*

Ex. 2. Given the square root of 160, equal to 12.64911;
" " 162, " 12.72792;
" " 164, " 12.80625,

to find the square root of 161.

Here the interval between the given numbers is 2; the distance of the required term from the first term is 1; and, since the interval of the given numbers is always to be reckoned as unity, we have $x=\frac{1}{2}$.

Also, $\text{D}'=+.07881$, $\text{D}''=-.00048$.
And $z=a+\frac{1}{2}\text{D}'-\frac{1}{8}\text{D}''$.

Therefore the square root of 161 is
12.64911+.03941+.00006,
or 12.68858. *Ans.*

Ex. 3. Given the cube root of 60, equal to 3.91487;
" " 62, " 3.95789;
" " 64, " 4.00000,
" " 66, " 4.04124,

to find the cube root of 61.

Ans. 3.93650.

Ex. 4. Given the fourth root of 625, equal to 5.000000;
" " 628, " 5.005988;
" " 631, " 5.011956;
" " 634, " 5.017903,

to find the fourth root of 627.

Here $x=\frac{2}{3}$. Therefore, $z=a+\frac{2}{3}\text{D}'-\frac{1}{9}\text{D}''$.

Ans. 5.003994

Ex. 5. Given the square root of 70, equal to 8.36660;

"	"	74,	"	8.60233;
"	"	78,	"	8.83176;
"	"	82,	"	9.05539,

to find the square root of 71.

Ans. 8.42615.

(298.) *Fractions expanded into infinite series.*

When the dividend is not exactly divisible by the divisor, the quotient may be expressed by a fraction. Thus, if it is required to divide 1 by $1-a$, we obtain the fraction $\frac{1}{1-a}$. We may, however, commence the division according to the usual method; thus,

$$\begin{array}{l|l} 1 & 1-a \\ \underline{1-a} & \overline{1+a+a^2+a^3+a^4+, \text{ \&c.}}, = \text{the quotient.} \\ \quad a & \\ \quad \underline{a-a^2} & \\ \qquad a^2 & \\ \qquad \underline{a^2-a^3} & \\ \qquad\quad a^3 & \\ \qquad\quad \underline{a^3-a^4} & \\ \qquad\qquad a^4 & \end{array}$$

Hence $\frac{1}{1-a}=1+a+a^2+a^3+a^4+a^5+$, &c., to infinity.

Suppose $a=\frac{1}{2}$, we shall then have

$\frac{1}{1-a}=\frac{1}{1-\frac{1}{2}}=2$, which will be equal to the series

$$1+\tfrac{1}{2}+\tfrac{1}{4}+\tfrac{1}{8}+\tfrac{1}{16}+, \text{ \&c.}$$

Suppose $a=\frac{1}{3}$, we shall then have

$\frac{1}{1-a}=\frac{1}{1-\frac{1}{3}}=\frac{3}{2}$, which will be equal to the series

$$1+\tfrac{1}{3}+\tfrac{1}{9}+\tfrac{1}{27}+\tfrac{1}{81}+, \text{ \&c.}$$

Ex. 2. Resolve $\frac{1}{1+a}$ into an infinite series.

Ans. $1-a+a^2-a^3+a^4-a^5+$, &c

Suppose $a=\frac{1}{2}$, we shall then have

$\frac{1}{1+\frac{1}{2}}=\frac{2}{3}$, which will be equal to the series

$$1-\tfrac{1}{2}+\tfrac{1}{4}-\tfrac{1}{8}+\tfrac{1}{16}-\tfrac{1}{32}+, \text{\&c.}$$

Ex. 3. Resolve the fraction $\frac{c}{a+b}$ into an infinite series.

Ans. $\frac{c}{a}-\frac{bc}{a^2}+\frac{b^2c}{a^3}-\frac{b^3c}{a^4}+$, &c.

Ex. 4. Resolve $\frac{a^2}{b+x}$ into an infinite series.

Ex. 5. Resolve $\frac{1+x}{1-x}$ into an infinite series.

We may proceed in the same manner when there are more than two terms in the divisor.

Ex. 6. Resolve $\frac{1}{1-a+a^2}$ into an infinite series.

Ans. $1+a-a^3-a^4+a^6+$, &c.

Ex. 7. Resolve $\frac{a^2}{(a+x)^2}$ into an infinite series.

(299.) *Infinite series obtained by extracting the square root.*

In *Art.* 272, $\sqrt{a+b}$ has been expanded into an infinite series by the Binomial Theorem. It was also remarked that the same result might have been obtained by extracting the square root according to the usual rule, *Art.* 275. The operation will proceed as follows:

$$a+b \,\Big|\, a^{\frac{1}{2}}+\frac{b}{2a^{\frac{1}{2}}}-\frac{b^2}{8a^{\frac{3}{2}}}+\frac{b^3}{16a^{\frac{5}{2}}}-, \text{\&c.}, = \text{the square root of } a+b$$

$$a$$

$$b \qquad \Big|\, 2a^{\frac{1}{2}}+\frac{b}{2a^{\frac{1}{2}}}, \text{ first divisor.}$$

$$b+\frac{b^2}{4a}$$

$$-\frac{b^2}{4a} \,\Big|\, 2a^{\frac{1}{2}}+\frac{b}{a^{\frac{1}{2}}}-\frac{b^2}{8a^{\frac{3}{2}}}, \text{ second divisor.}$$

$$-\frac{b^2}{4a}-\frac{b^3}{8a^2}+\frac{b^4}{64a^3}$$

$$+\frac{b^3}{8a^2}-\frac{b^4}{64a^3}.$$

This result is the same as that obtained in *Art.* 272.

Ex. 2. Extract the square root of $1-x$.

$$Ans.\ 1-\frac{x}{2}-\frac{x^2}{8}-\frac{x^3}{16}-\frac{5x^4}{128}-, \&c.$$

Ex. 3. Extract the square root of a^2+b.

Ex. 4. Extract the square root of a^4-b.

METHOD OF UNKNOWN COEFFICIENTS.

(300.) The method of unknown coefficients is a method of developing algebraic expressions into series, by assuming a series having unknown coefficients, and afterward finding the value of these coefficients. This method is founded on the following

THEOREM.

If an equation of the form

$$A+Bx+Cx^2+Dx^3+, \&c., =A'+B'x+C'x^2+D'x^3+, \&c.,$$

must be verified by any value given to x, *the terms involving the same powers in the two members are respectively equal.*

For, since this equation must be verified for every value of x, it must be verified when $x=0$. But, upon this supposition, all the terms vanish except two, and we have

$$A=A'.$$

Suppressing these two equal terms, we have

$$Bx+Cx^2+Dx^3+, \&c., =B'x+C'x^2+D'x^3+, \&c.$$

Dividing every term by x, we obtain

$$B+Cx+Dx^2+, \&c., =B'+C'x+D'x^2+, \&c.$$

Since this equation must be verified for every value of x, it must be verified when $x=0$. But, upon this supposition,

$$B=B'.$$

In the same manner, we can prove that

$$C=C',$$
$$D=D', \&c.$$

(301.) Let it be proposed to develop the expression $\frac{1-x}{1+x}$ into a series arranged according to the powers of x. It is plain that this development is possible, for we may divide the numerator by the denominator, as explained in *Art.* 298.

Let us, then, assume

$$\frac{1-x}{1+x}=A+Bx+Cx^2+Dx^3+Ex^4+, \&c.,$$

where the coefficients A, B, C, D are supposed to be independent of x, but dependent on the known terms of the fraction.

In order to obtain the values of these coefficients, let us multiply both members of the above equation by the denominator $1+x$, and we shall have

$$1-x=A+(A+B)x+(B+C)x^2+(C+D)x^3+(D+E)x^4+, \&c.$$

But, according to the preceding Theorem, the terms involving the same powers of x in the two members of the equation must be equal to each other.

Therefore, $A=1$,

$A+B=-1$; hence $B=-2$.

$B+C=0$; " $C=+2$.

$C+D=0$; " $D=-2$.

$D+E=0$; " $E=+2$.

&c., &c.

Substituting these values of the coefficients in the assumed series, we obtain

$$\frac{1-x}{1+x}=1-2x+2x^2-2x^3+2x^4-, \&c.$$

(302.) The method thus exemplified is expressed in the following

RULE.

Assume a series with unknown coefficients as equal to the proposed expression; then, having cleared the equation of fractions, or raised it to its proper power, find the value of each of these coefficients by equating the corresponding terms of the two expressions, or putting such of them as have no corresponding terms, equal to zero.

Ex. 2. Expand the fraction $\frac{1}{1-2x+x^2}$ into an infinite series

Assume $\frac{1}{1-2x+x^2}=A+Bx+Cx^2+Dx^3+Ex^4+, \&c.$

Multiplying by $1-2x+x^2$, we have

$1=A+(B-2A)x+(C-2B+A)x^2+(D-2C+B)x^3+(E-2D+C)x^4+$, &c.

Hence we must have

$$\begin{aligned} A&=1 \\ B-2A&=0 \therefore B=2A \quad =2, \\ C-2B+A&=0 \therefore C=2B-A=3, \\ D-2C+B&=0 \therefore D=2C-B=4, \\ E-2D+C&=0 \therefore E=2D-C=5, \\ &\&c., \qquad \&c. \end{aligned}$$

Therefore, $\frac{1}{1-2x+x^2}=1+2x+3x^2+4x^3+5x^4+$, &c.

Ex. 3. Expand the fraction $\frac{1+2x}{1-x-x^2}$ into an infinite series.

Ans. $1+3x+4x^2+7x^3+11x^4+18x^5+29x^6+$, &c.,

where the coefficient of each term is equal to the sum of the coefficients of the two preceding terms.

Ex. 4. Expand $\frac{1-x}{1-2x-3x^2}$ into an infinite series.

Ans. $1+x+5x^2+13x^3+41x^4+121x^5+$, &c

What is the law of the coefficients in this series?

Ex. 5. Expand $\frac{1+2x}{1-3x}$ into an infinite series.

Ans. $1+5x+15x^2+45x^3+135x^4+$, &c.

What is the law of the coefficients in this series?

Ex. 6. Expand $\sqrt{1-x}$ into an infinite series.

Ans. $1-\frac{x}{2}-\frac{x^2}{2.4}-\frac{3x^3}{2.4.6}-\frac{3.5x^4}{2.4.6.8}-\frac{3.5.7x^5}{2.4.6.8.10}-$, &c.

(303.) The method of unknown coefficients requires that we should know beforehand the *form* of the development with respect to the powers of x. Generally, we suppose the development to proceed according to the ascending powers of x, commencing with x^0; but sometimes this form is *inapplicable*, in which case the result of the operation is sure to indicate it.

Let it be required, for example, to develop the expression $\frac{1}{3x-x^2}$ into a series.

Assume $\qquad \frac{1}{3x-x^2}=A+Bx+Cx^2+Dx^3+$, &c.

Clearing of fractions, we have

$$1=3Ax+(3B-A)x^2+(3C-B)x^3+, \&c.;$$

whence, according to *Art.* 300, we conclude

$$1=0,$$
$$3A=0, \&c.$$

Now the first equation, $1=0$, is *absurd*, and shows that the assumed form is not applicable in the present case. But if we put the fraction under the form $\frac{1}{x}\times\frac{1}{3-x}$, and suppose that

$$\frac{1}{x}\times\frac{1}{3-x}=\frac{1}{x}(A+Bx+Cx^2+Dx^3+, \&c.),$$

it will become, after the reductions are made,

$$1=3A+(3B-A)x+(3C-B)x^2+(3D-C)x^3+, \&c.,$$

which gives the equations

$$3A=1; \text{ whence } A=\tfrac{1}{3}.$$
$$3B-A=0; \quad \text{"} \quad B=\tfrac{1}{9}.$$
$$3C-B=0; \quad \text{"} \quad C=\tfrac{1}{27}.$$
$$3D-C=0; \quad \text{"} \quad D=\tfrac{1}{81}.$$

Therefore, $\frac{1}{3x-x^2}=\frac{1}{x}\left(\frac{1}{3}+\frac{x}{9}+\frac{x^2}{27}+\frac{x^3}{81}+, \&c.\right)$

$$=\frac{x^{-1}}{3}+\frac{x^0}{9}+\frac{x}{27}+\frac{x^2}{81}+, \&c.;$$

that is, the development contains a term affected with a *negative exponent.*

We ought, then, to have assumed at the outset

$$\frac{1}{3x-x^2}=Ax^{-1}+B+Cx+Dx^2+Ex^3+, \&c.$$

The particular series which should be adopted in each case may be determined by putting $x=0$, and observing the nature of the result. If, in this case, the proposed expression becomes equal to a finite quantity, the first term of the series will not contain x. If the expression reduces to zero, the first term will contain x; and if the expression reduces to the form $\frac{A}{0}$, then the first term of the development must contain x with a negative exponent.

SECTION XIX.

GENERAL THEORY OF EQUATIONS.

(304.) It is proposed in this Section to exhibit the most important propositions relating to the theory of equations, together with the Theorem of Sturm, by which we are enabled to determine the number of real roots of an equation.

A *function* of a quantity is any expression involving that quantity. Thus,

ax^2+b is a function of x.
ay^3+cy+d is a function of y.
ax^2-by^2 is a function of x and y.

In a series of terms, two successive signs constitute a *permanence* when the signs are alike, and a *variation* when they are unlike. Thus, in the polynomial

$$a+b-c+d,$$

the signs of the first two terms constitute a permanence; the signs of the second and third constitute a variation: and those of the third and fourth also a variation.

(305.) A *cubic equation* is one in which the highest power of the unknown quantity is of the third degree; as, for example,

$$x^3-6x^2+8x-15=0.$$

All equations of the third degree, with one unknown quantity, may be reduced to the form

$$x^3+ax^2+bx+c=0.$$

A *biquadratic equation* is one in which the highest power of the unknown quantity is of the fourth degree; as, for example,

$$x^4-6x^3+7x^2+5x-4=0.$$

Every equation of the fourth degree, with one unknown quantity, may be reduced to the form

$$x^4+ax^3+bx^2+cx+d=0.$$

The general form of an equation of the fifth degree is

$$x^5+ax^4+bx^3+cx^2+dx+e=0;$$

and the general form of an equation of the mth degree, with one unknown quantity, is

$$x^m+Ax^{m-1}+Bx^{m-2}+Cx^{m-3}+\ldots\ldots+Tx+V=0 \quad (m).$$

This equation will be frequently referred to hereafter by the name of *the general equation of the* m*th degree*, or simply by the letter (m).

It is obvious, that if we could solve this equation, we should have the solution of every equation which could be proposed. Unfortunately, no general solution has ever been discovered; yet many important properties are known, which enable us to solve any numerical equation which can ever occur.

PROPOSITION I.

(306.) *If* a *is a root of the general equation of the* m*th degree, the equation will be exactly divisible by* x−a.

For if a is one value of x, the equation must be verified when we substitute a in the place of x. Hence we must have

$$a^m+Aa^{m-1}+Ba^{m-2}+Ca^{m-3}+\ldots\ldots+Ta+V=0 \quad (1).$$

Subtracting equation (1) from equation (m), we obtain

$$(x^m-a^m)+A(x^{m-1}-a^{m-1})+B(x^{m-2}-a^{m-2})+\ldots+T(x-a)=0 \quad (2).$$

But, by *Art.* 76, each of the expressions (x^m-a^m), $(x^{m-1}-a^{m-1})$, &c., is divisible by $x-a$, and therefore equation (2) is also divisible by $x-a$. Now equation (m) is but another form for equation (2); for if we take the value of V, as found from equation (1), and substitute it for V in equation (m), it will give us equation (2); therefore, equation (m) is divisible by $x-a$.

Conversely, if equation (m) is divisible by $x-a$, then a is a root of the equation.

It will be noticed that this property is but a *generalization* of what has been proved of equations of the second degree, in *Art.* 192.

Ex. 1. Prove that 1 is a root of the equation

$$x^3-6x^2+11x-6=0.$$

This equation is divisible by $x-1$, and gives $x^2-5x+6=0$

Ex. 2. Prove that 2 is a root of the equation

$$x^3-x-6=0.$$

This equation is divisible by $x-2$, and gives $x^2+2x+3=0$.

Ex. 3. Prove that 2 is a root of the equation

$$x^3-11x^2+36x-36=0.$$

Ex. 4. Prove that 4 is a root of the equation

$$x^3+x^2-34x+56=0.$$

Ex. 5. Prove that -1 is a root of the equation

$$x^4-38x^3+210x^2+538x+289=0.$$

Ex. 6. Prove that -5 is a root of the equation

$$x^5+6x^4-10x^3-112x^2-207x-110=0.$$

Ex. 7. Prove that 3 is a root of the equation

$$x^7+x^6-14x^5-14x^4+49x^3+49x^2-36x-36=0.$$

PROPOSITION II.

(307.) *Every equation of the* m*th degree containing but one unknown quantity, has* m *roots and no more.*

For, suppose a to be a root of the general equation of the mth degree. By the last Proposition, this equation is divisible by $x-a$; and if we actually perform the division, the equation will be reduced to one of the next inferior degree.

If we represent the coefficients of the different powers of x by A′, B′, &c., the quotient will be

$$x^{m-1}+A'x^{m-2}+B'x^{m-3}+\ldots\ldots+T'x+V'=0.$$

This equation must also have a root, which we will represent by b; and dividing by $x-b$, the equation will be reduced to one of the next inferior degree, and so on.

We may continue this series of operations $(m-1)$ times, when we shall arrive at a simple equation which has only one root. Hence the proposed equation will have m roots,

$$a, b, c, d, \ldots\ldots l;$$

and its successive divisors, or the factors of which it is composed, will be

$$x-a, x-b, x-c, x-d, \ldots\ldots x-l,$$

being equal in number to the units contained in m, the highest exponent of the equation.

We have seen that when one root of an equation is known, the equation is readily reduced to one of the next inferior degree; and if we can depress any equation to a quadratic, its roots can be determined by methods already explained.

Ex. 1. One root of the equation

$$x^3+3x^2-16x+12=0$$

is 1. Find the remaining roots.

Ex. 2. Two roots of the equation

$$x^4-10x^3+35x^2-50x+24=0$$

are 1 and 3. Find the remaining roots.

Ex. 3. Two roots of the equation

$$x^4-12x^3+48x^2-68x+15=0$$

are 3 and 5. Find the remaining roots.

Ans. $2\pm\sqrt{3}$.

Ex. 4. Two roots of the equation

$$4x^4-14x^3-5x^2+31x+6=0$$

are 2 and 3. Find the remaining roots.

Ans. $\frac{-3\pm\sqrt{5}}{4}$.

Ex. 5. Two roots of the equation

$$x^4-6x^3+24x-16=0$$

are 2 and -2. Find the remaining roots.

Ans. $3\pm\sqrt{5}$.

(308.) It should be observed that this Proposition only proves that an equation of the mth degree may be continually depressed by division, and finally exhausted after m operations. The divisors are not necessarily *unequal. Any number, and indeed all of them, may be equal.* When we say that an equation of the mth degree has m roots, we mean that the polynomial can be decomposed into m binomial factors, equal or unequal, each containing one root. Thus, the equation

$$x^3-6x^2+12x-8=0$$

can be resolved into the factors

$$(x-2)\ (x-2)\ (x-2)=0;\ \text{or}\ (x-2)^3=0;$$

whence it appears that the three roots of this equation are

2, 2, 2.

But, in general, the several roots of an equation differ from each other numerically.

The equation

$$x^3=8$$

has apparently but one root, viz., 2; but by the method of the preceding article we can discover two other roots. Dividing x^3-8 by $x-2$, we obtain

$$x^2+2x+4=0.$$

Solving this equation, we find

$$x=-1\pm\sqrt{-3}.$$

Thus, the three roots of the equation $x^3=8$ are

$$2;\quad -1+\sqrt{-3};\quad -1-\sqrt{-3}.$$

These last two values may be verified by multiplication as follows:

$$\begin{array}{rl}
-1+\sqrt{-3} & \\
-1+\sqrt{-3} & \\
\hline
1-\sqrt{-3} & \\
-\sqrt{-3}-3 & \\
\hline
-2-2\sqrt{-3} & =\text{the square.} \\
-1+\sqrt{-3} & \\
\hline
2+2\sqrt{-3} & \\
-2\sqrt{-3}+6 & \\
\hline
8 & =\text{the cube.}
\end{array}
\qquad
\begin{array}{rl}
-1-\sqrt{-3} & \\
-1-\sqrt{-3} & \\
\hline
1+\sqrt{-3} & \\
+\sqrt{-3}-3 & \\
\hline
-2+2\sqrt{-3} & =\text{the square.} \\
-1-\sqrt{-3} & \\
\hline
2-2\sqrt{-3} & \\
+2\sqrt{-3}+6 & \\
\hline
8 & =\text{the cube.}
\end{array}$$

If the last term of an equation vanishes, as in the example

$$x^4+2x^3+3x^2+6x=0,$$

the equation is divisible by $x-0$, and, consequently, 0 is one of its roots.

If the last two terms vanish, then two of its roots are equal to 0.

PROPOSITION III.

To discover the law of the coefficients of every equation.

(309.) In order to discover the law of the coefficients, let us form the equation whose roots are

$$a,\ b,\ c,\ d, \ldots\ldots l.$$

This equation will contain the factors $(x-a)$, $(x-b)$, $(x-c)$, &c.; that is, we shall have

$$(x-a)\ (x-b)\ (x-c)\ (x-d)\ \ldots\ldots\ (x-l)=0.$$

If we perform the multiplication as in *Art.* 264, we shall have

$$\begin{array}{l|l|l|l}
x^m - a & x^{m-1} + ab & x^{m-2} - abc & x^{m-3} + \ldots\ldots - (abc\ldots\ldots l) = 0. \\
\quad - b & \quad + ac & \quad - abd & \\
\quad - c & \quad + ad & \quad - acd & \\
\quad - d & \quad + bc & \quad - bcd & \\
\quad \&c. & \quad + bd & \quad \&c. & \\
 & \quad + cd & & \\
 & \quad \&c. & &
\end{array}$$

Hence we perceive,

1. *The coefficient of the second term of any equation is equal to the sum of all the roots with their signs changed.*

2. *The coefficient of the third term is equal to the sum of the products of all the roots taken two and two.*

3. *The coefficient of the fourth term is equal to the sum of the products of all the roots taken three and three, with their signs changed.*

4. *The last term is the product of all the roots with their signs changed.*

It will be perceived that these properties include those of quadratic equations mentioned on pages 163 and 164.

If the roots are all negative, the signs of all the terms of the equation will be positive, because the factors of which the equation is composed are all positive.

If the roots are all positive, the signs of the terms will be alternately + and −.

Ex. 1. Form the equation whose roots are 1, 2, and 3.

For this purpose, we must multiply together the factors $x-1$, $x-2$, $x-3$, and we obtain

$$x^3-6x^2+11x-6=0.$$

This example conforms to the rules above given for the coefficients. Thus, the coefficient of the second term is equal to the sum of all the roots $(1+2+3)$ with their signs changed.

The coefficient of the third term is the sum of the products of the roots taken two and two ; thus,

$$1\times2+1\times3+2\times3=11.$$

The last term is the product of all the roots $(1\times2\times3)$ with their signs changed.

Ex. 2. Form the equation whose roots are 2, 3, 5, and -6.

Ans. $x^4-4x^3-29x^2+156x-180=0$.

Show how these coefficients conform to the laws above given.

Ex. 3. Form the equation whose roots are 1, 3, 5, -2, -4, -6.

Ans. $x^6+3x^5-41x^4-87x^3+400x^2-444x-720=0$.

(310.) *Every rational root of an equation is a divisor of the last term;* for, since this term is the product of all the roots, it must be *divisible* by each of them. If, then, we wish to find a root by trial, we know at once what numbers we must employ.

For example, take the equation

$$x^3-x-6=0.$$

If this equation has a rational root, it must be a divisor of the last term, 6; hence we must try the numbers 1, 2, 3, 6, either positive or negative.

If	$x=1$, we have	$1-1-6=-6$,
	$x=2$, "	$8-2-6=\ 0$,
	$x=3$, "	$27-3-6=18$,
	$x=6$, "	$216-6-6=204$,

Hence we see that 2 is one of the roots of the given equation, and by the method of *Art.* 307, we shall find the remaining roots to be

$$-1\pm\sqrt{-2}.$$

PROPOSITION IV.

(311.) *No equation whose coefficients are all integers, and that of the first term unity, can have a root equal to a rational fraction.*

For, take the general equation of the third degree,

$$x^3+Ax^2+Bx+C=0,$$

and suppose, if possible, that the fraction $\frac{a}{b}$ is one value of x, this fraction being reduced to its lowest terms. If we substitute this value for x in the given equation, we shall have

$$\frac{a^3}{b^3}+A\frac{a^2}{b^2}+B\frac{a}{b}+C=0.$$

Multiplying each term by b^2, and transposing, we obtain

$$\frac{a^3}{b}=-(Aa^2+Bab+Cb^2).$$

Now, by supposition, A, B, C, a and b are whole numbers. Hence the entire right-hand member of the equation is a whole number.

But by hypothesis, $\frac{a}{b}$ is an irreducible fraction; that is, a and b contain no common factor. Consequently, a^3 and b will contain no common factor, that is, $\frac{a^3}{b}$ is a fraction in its lowest terms. Hence, if $\frac{a}{b}$ were a root of the proposed equation, we should have a fraction in its lowest terms equal to a whole number, which is *absurd.*

The same mode of demonstration is applicable to the general equation of the mth degree.

This proposition only asserts that in an equation such as is here described, the real roots must be integers, or they can not be *exactly* expressed in numbers. They may often be expressed *approximately* by fractions, as is seen in the examples on pages 288–301. A real root which can not be exactly expressed in numbers is called *incommensurable.*

PROPOSITION V.

(312.) *If the signs of the alternate terms in an equation are changed, the signs of all the roots will be changed.*

If we take the general equation of the mth degree, and change the signs of the alternate terms, we shall have

$$x^m-Ax^{m-1}+Bx^{m-2}-Cx^{m-3}+\ldots\ldots=0 \quad (1):$$

or, changing the sign of every term of the last equation,

$$-x^m+Ax^{m-1}-Bx^{m-2}+Cx^{m-3}-\ldots\ldots=0 \quad (2).$$

Now, substituting $+a$ for x in equation (m) will give the same result as substituting $-a$ in equation (1), if m be an *even* number; or, substituting $-a$ in equation (2), if m be an *odd* number. If, then, a is a root of equation (m), $-a$ will be a root of equation (1), and, of course, a root of equation (2), which is identical with it.

Hence we see that the positive roots may be changed into negative roots, and the reverse, by simply changing the signs of the alternate terms; so that the finding the real roots of any equation is reduced to finding positive roots only.

Ex. 1. The roots of the equation

$$x^3-2x^2-5x+6=0$$

are 1, 3, and -2. What are the roots of the equation

$$x^3+2x^2-5x-6=0\,?$$

Ex. 2. The roots of the equation

$$x^3-6x^2+11x-6=0$$

are 1, 2, and 3. What are the roots of the equation

$$x^3+6x^2+11x+6=0\,?$$

PROPOSITION VI.

(313.) *If an equation whose coefficients are all real, contains imaginary roots, the number of these roots must be even.*

If an equation whose coefficients are all real, has a root of the form

$$a+b\sqrt{-1},$$

then will

$$a-b\sqrt{-1}$$

be also a root of the equation.

For, let $a+b\sqrt{-1}$ be substituted for x in the equation, the result will consist of a series of terms, of which those involving only the powers of a, and the even powers of $b\sqrt{-1}$ will be real, and those which involve the odd powers of $b\sqrt{-1}$ will be imaginary. If we denote the sum of the real terms by P, and the sum of the imaginary terms by $Q\sqrt{-1}$, then we must have

$$P+Q\sqrt{-1}=0,$$

which relation can only exist when $P=0$ and $Q=0$.

Again, let $a-b\sqrt{-1}$ be substituted for x in the proposed equation, the only difference in the result will be in the signs of the odd powers of $b\sqrt{-1}$, so that the result will be $P-Q\sqrt{-1}$. But we have found that $P=0$ and $Q=0$; hence

$$P-Q\sqrt{-1}=0.$$

And, since $a-b\sqrt{-1}$ substituted for x gives a result equal to zero, it must be a root of the equation.

Ex. 1. Find the roots of the equation

$$x^3-2x+4=0.$$

Ans. -2, and $1\pm\sqrt{-1}$.

Ex. 2. Find the roots of the equation

$$x^3-x^2-7x+15=0.$$

Ans. -3, and $2\pm\sqrt{-1}$.

Ex. 3. Find the roots of the equation

$$5x^3+2x-44=0.$$

Ans. 2, and $-1\pm\sqrt{-3.4}$.

Hence every equation of the third degree whose coefficients are all real, must have one real root. The same is true of every equation of an odd degree.

PROPOSITION VII.

(314.) *Every equation must have as many variations of sign as it has positive roots, and as many permanences of sign as there are negative roots.*

To prove this Proposition, it is only necessary to show that the multiplication of an equation by a new factor, $x-a$, corresponding to a *positive* root, will introduce at least *one variation*, and that the multiplication by a factor $x+a$ will introduce at least *one permanence.*

For an example, take the equation

$$x^3+3x^2-10x-24=0,$$

n which the signs are $++--$, giving *one variation.*

Multiply this equation by $x-2=0$, as follows:

$$\begin{array}{l} x^3+3x^2-10x-24 \\ x-2 \\ \hline x^4+3x^3-10x^2-24x \\ \quad\;\; -2x^3-\;\;6x^2+20x+48 \\ \hline x^4+\;x^3-16x^2-\;\;4x+48=0. \end{array}$$

In this last product the signs are $++--+$, giving *two variations;* that is, the introduction of a positive root has introduced one new variation in the signs of the terms.

To generalize this reasoning, we perceive that the signs in the upper line of the partial products are the *same* as in the given equation; but those in the lower line are all *contrary* to those of the given equation, and advanced one term toward the right.

Now, if each coefficient of the upper line is greater than the corresponding one in the lower, the signs of the upper line will be the same as in the total product, with the exception of the last term. But the last term introduces a new variation, since its sign is contrary to that which immediately precedes it; that is, the product contains one more variation than the original equation.

When a term in the lower line is larger than the corresponding one in the upper line, and has the contrary sign, there is a change from a permanence to a variation; for the lower sign is always contrary to the preceding upper sign. Hence, whenever we are obliged to descend from the upper to the lower line in order to determine the sign of the product, there is a variation which is not found in the proposed equation; and as all the remaining signs of the lower line are contrary to those of the proposed equation, there must be the same *changes* of sign in this line as in the given equation. If we are obliged to reascend to the upper line, the result may be either a variation or a permanence. But even if it were a permanence, since the last sign of the product is in the lower line, it is necessary to go once more from the upper line to the lower, than from the lower to the upper. Hence each factor, corresponding to a positive root, must introduce at least *one new variation;* so that there must be as many variations as there are positive roots.

In the same manner, we may prove that the multiplication by a factor $x+a$, corresponding to a *negative* root, must introduce at least one new *permanence;* so that there must be as many permanences as there are negative roots.

Ex. 1. The roots of the equation

$$x^5-3x^4-5x^3+15x^2+4x-12=0$$

are 1, 2, 3, -1, and -2. There are also three variations of sign, and two permanences, as there should be, according to the Proposition.

Ex. 2. The equation

$$x^4-3x^3-15x^2+49x-12=0$$

has four real roots. How many of these are negative?

Ex. 3. The equation

$$x^6+3x^5-41x^4-87x^3+400x^2+444x-720=0$$

has six real roots. How many of these are positive?

If all the roots of an equation are *real*, the number of positive roots must be the *same* as the number of variations, and the number of negative roots must be the *same* as the number of permanences. If any term of an equation is wanting, we must supply its place with ± 0 before applying the preceding Rule.

PROPOSITION VIII.

(315.) *If two numbers, when substituted for the unknown quantity in an equation, give results with contrary signs, there is at least one root comprised between those numbers.*

Take, for example, the equation

$$x^3-2x^2+3x-44=0.$$

If we substitute 3 for x in this equation, we obtain -26; and if we substitute 5 for x, we obtain $+46$. There must, therefore, be a real root between 3 and 5; for, when we suppose $x=3$, we have

$$x^3+3x<2x^2+44.$$

But when we suppose $x=5$, we have

$$x^3+3x>2x^2+44.$$

Now both the quantities

$$x^3+3x \text{ and } 2x^2+44$$

increase while x increases. And since the first of these quantities, which was originally less than the second, has become the greater, it must increase more rapidly than the second. There must, therefore, be a point at which the two magnitudes are equal, and that value of x which renders these two magnitudes equal must be a root of the proposed equation.

In general, if two numbers, p and q, substituted for x in an equation, give results with contrary signs, we may suppose the less of the two numbers to increase by imperceptible degrees until it becomes equal to the greater number. The results of these successive substitutions must also change by imperceptible degrees, and must pass through all the intermediate values between the two extremes. But the two extreme values are affected with opposite signs; there must, therefore, be some number between p and q which reduces the given equation to zero, and this number will be a root of the equation.

In the same manner, it may be proved that *if any quantity* p, *and every quantity greater than* p, *substituted in an equation, renders the result positive, then* p *is greater than the greatest root.*

Hence, also, if the signs of the alternate terms are changed, and *if* q, *and every quantity greater than* q, *renders the result positive, then* $-$q *is less than the least root.*

If the two numbers, which give results with contrary signs, differ from each other only by unity, it is plain that we have found the *integral part* of a root.

Ex. 1. Find the integral part of one of the roots of the equation

$$2x^4-11x^2+8x-16=0.$$

When $x=2$, the equation reduces to -12; and when $x=3$, it reduces to $+71$. Hence there must be a root between 2 and 3; that is, 2 is the first figure of one of the roots.

Ex. 2. Find the first figure of one of the roots of the equation

$$x^3+x^2+x-100=0.$$

Ans. 4.

Ex. 3. Find the first figure of each of the roots of the equation

$$x^3-4x^2-6x+8=0.$$

PROPOSITION IX.

(316.) *Every equation may be transformed into another, whose roots are greater or less than those of the former by any given quantity.*

Let it be required to transform the general equation of the mth degree into another whose roots are greater by r than those of the given equation.

Take $y=x+r$, or $x=y-r$,

and substitute $y-r$ for x in the proposed equation; we shall then have

$$y^m \left.\begin{matrix} -mr \\ +A \end{matrix}\right| y^{m-1} + \left.\begin{matrix} \frac{m(m-1)r^2}{2} \\ -(m-1)Ar \\ +B \end{matrix}\right| y^{m-2} \left.\begin{matrix} -\frac{m(m-1)(m-2)r^3}{2.3} \\ +\frac{(m-1)(m-2)Ar^2}{2} \\ -(m-2)Br \\ +C \end{matrix}\right| y^{m-3}, \&c., =0,$$

which equation evidently fulfills the required conditions, since y is greater than x by r.

If we take $y=x-r$, or $x=y+r$, we shall obtain in the same way an equation whose roots are *less* than those of the given equation by r.

Ex. 1. Find the equation whose roots are greater by 1 than those of the equation

$$x^3+3x^2-4x+1=0.$$

We must here substitute $y-1$ in place of x.

Ans. $y^3-7y+7=0.$

Ex. 2. Find the equation whose roots are less by 1 than those of the equation

$$x^3-2x^2+3x-4=0.$$

Ans. $y^3+y^2+2y-2=0.$

Ex. 3. Find the equation whose roots are greater by 3 than those of the equation

$$x^4+9x^3+12x^2-14x=0.$$

Ans. $y^4-3y^3-15y^2+49y-12=0.$

Ex. 4. Find the equation whose roots are less by 2 than those of the equation

$$5x^4-12x^3+3x^2+4x-5=0.$$

Ans. $5y^4+28y^3+51y^2+32y-1=0.$

Ex. 5. Find the equation whose roots are greater by 2 than those of the equation

$$x^5+10x^4+42x^3+86x^2+70x+12=0.$$

Ans. $y^5+2y^3-6y^2-10y+8=0.$

PROPOSITION X.

(317.) *Any complete equation may be transformed into another whose second term is wanting.*

Since r in the preceding Proposition is indeterminate, we may put $-mr+A$ equal to zero, which will cause the second term of the general development to disappear. Hence $r=\frac{A}{m}$, and $x=y-\frac{A}{m}$.

Hence, to remove the second term of an equation, *substitute for the unknown quantity a new unknown quantity, together with such a part of the coefficient of the second term, taken with a contrary sign, as is denoted by the degree of the equation.*

Ex. 1. Transform the equation

$$x^3-6x^2+8x-2=0$$

into another whose second term is wanting.

Here we take a new unknown quantity, and annex to it a third part of the coefficient of the second term of the equation with its sign changed; that is, we put $x=y+2$. Making this substitution, we obtain

$$y^3-4y-2=0. \quad \textit{Ans.}$$

Ex. 2. Transform the equation

$$x^4-16x^3-6x+15=0$$

into another whose second term is wanting.

Here we put $x=y+4$.

$$\textit{Ans.} \quad y^4-96y^2-518y-777=0$$

Ex. 3. Transform the equation

$$x^5+15x^4+12x^3-20x^2+14x-25=0$$

into another whose second term is wanting.

$$\textit{Ans.} \quad y^5-78y^3+412y^2-757y+401=0.$$

Since the coefficient of the second term is equal to the sum of the roots with their signs changed, it is obvious that when the second term of an equation is wanting, the sum of the positive roots must be equal to the sum of the negative roots

PROPOSITION XI.

(318.) *To discover the law of Derived Polynomials.*

When we substitute $y+r$ for x in the general equation of the mth degree, the coefficients of r follow a remarkable law

The equation, before it is developed, is

$$(y+r)^m+A(y+r)^{m-1}+B(y+r)^{m-2}+\ldots\ldots+T(y+r)+V=0.$$

If we actually involve the several terms $(y+r)^m$, $(y+r)^{m-1}$, &c., as was done in *Art.* **316**, we obtain certain terms independent of r, others which contain the first power of r, others the second power of r, and so on; and the development is of the following form:

$$X+X_1r+\frac{X_2}{2}r^2+\frac{X_3}{2.3}r^3+\frac{X_4}{2.3.4}r^4+\ldots\ldots+r^m,$$

where the values of X, X_1, X_2, &c., are

$$X=y^m+Ay^{m-1}+By^{m-2}+Cy^{m-3}+\qquad\ldots\ldots+Ty+V.$$
$$X_1=my^{m-1}+(m-1)Ay^{m-2}+(m-2)By^{m-3}+\ldots\ldots+2Sy+T.$$
$$X_2=m(m-1)y^{m-2}+(m-1)(m-2)Ay^{m-3}+\ldots\ldots$$

Each of these polynomials may be derived from that immediately preceding it, by multiplying each term by the exponent of y *in that term, and diminishing the exponent by unity.*

The expressions X_1, X_2, &c., are called derived polynomials of X. X_1 is called the *first derived polynomial*, X_2 the *second derived polynomial*, X_3 the *third*, and so on.

Ex. 1. Find the equation whose roots are less by r than those of the equation

$$x^2-5x+6=0.$$

Here we shall have

$$X=y^2-5y+6,$$
$$X_1=2y-5,$$
$$X_2=2,$$
$$X_3=0.$$

But we have seen that when $y+r$ is substituted for x, the equation reduces to the form

$$X+X_1r+\frac{X_2}{2}r^2+\frac{X_3}{2.3}r^3+,\ \&c.$$

Substituting the values of X, X_1, X_2, &c., above found, we obtain

$$(y^2-5y+6)+(2y-5)r+r^2,$$

which is the development of

$$(y+r)^2-5(y+r)+6.$$

Ex. 2. Find the equation whose roots are less by r than those of the equation

$$x^3-7x^2+8x-3=0.$$

Here we shall have

$$\begin{aligned} X &= y^3-7y^2+8y-3, \\ X_1 &= 3y^2-14y+8, \\ X_2 &= 6y-14, \\ X_3 &= 6, \\ X_4 &= 0; \end{aligned}$$

and, substituting these values in the same formula as above, we obtain

$$(y^3-7y^2+8y-3)+(3y^2-14y+8)r+\frac{1}{2}(6y-14)r^2+\frac{1}{2.3}6r^3,$$

which is the development of

$$(y+r)^3-7(y+r)^2+8(y+r)-3.$$

Ex. 3. Find the successive derived polynomials of the equation

$$x^4-8x^3+14x^2+4x-8=0.$$

Ex. 4. Find the successive derived polynomials of the equation

$$x^5+3x^4+2x^3-3x^2-2x-2=0.$$

PROPOSITION XII.

(319.) *To find the equal roots of an equation.*

We have seen, in *Art.* 308, that an equation may have two or more *equal roots.* Thus, the equation

$$x^3-6x^2+12x-8=0,$$

or

$$(x-2)^3=0,$$

has the three equal roots 2, 2, 2. Such an equation and its first derived polynomial always contain a *common divisor;* for the first derived polynomial of the above equation is

$$3x^2-12x+12,$$

or

$$3(x-2)^2,$$

where it is evident that $(x-2)^2$ is a common divisor of both equations.

In general, let a be one of the equal roots which occurs n times as a root of the given equation; the first member will therefore contain the factors $(x-a)$, $(x-a)$, $(x-a)$, – – – ; that is, $(x-a)^n$. The first derived polynomial will contain the factor $n(x-a)^{n-1}$; that is, $x-a$ occurs $(n-1)$ times as a factor in the first derived polynomial. The greatest common divisor of the given equation and its first derived polynomial must therefore contain the factor $(x-a)$ repeated once less than in the given equation.

To determine, therefore, whether an equation has equal roots, *find the greatest common divisor between the equation and its first derived polynomial. If there is no common divisor, the equation has no equal roots. If there is a common divisor, solve the equation obtained by putting this divisor equal to zero.*

Ex. 1. Find the equal roots of the equation

$$x^3-8x^2+21x-18=0.$$

The first derived polynomial of this equation is

$$3x^2-16x+21.$$

The greatest common divisor between this and the given equation is

$$x-3.$$

Hence the equation has two roots, each equal to 3.

Ex. 2. Find the equal roots of the equation

$$x^3-13x^2+55x-75=0.$$

Ans. Two roots equal to 5.

Ex. 3. Find the equal roots of the equation

$$x^3-7x^2+16x-12=0.$$

Ans. Two roots equal to 2.

Ex. 4. Find the equal roots of the equation

$$x^4-6x^2-8x-3=0.$$

Ans. Three roots equal to -1

PROPOSITION XIII.

(320.) *To find the number of real and imaginary roots of an equation.*

In 1829, M. Sturm discovered a theorem which determines the precise number of real roots, and of course the number of imaginary ones, since the real and imaginary roots are together equal in number to the degree of the equation. We propose now to develop this theorem.

Let X represent the first member of the general equation of the mth degree, which we suppose to have no equal roots, and let $X_{,}$ be its first derived polynomial, found by the method of *Art.* 318.

Divide X by $X_{,}$ until the remainder is of a lower degree than the divisor, and call this remainder $-X_{,,}$; that is, let $X_{,,}$ designate the remainder with a contrary sign. Divide $X_{,}$ by $X_{,,}$ in the same manner, and so on, designating the successive remainders with contrary signs by $X_{,,,}$, $X_{,,,,}$, &c., until the division terminates by leaving a numerical remainder independent of x; which must always be the case, according to the preceding Proposition, since the equation having no equal roots, there can be no factor, which is a function of x, common to the equation and its first derived polynomial. Let this remainder, having its signs changed, be called X_{m}.

The operation thus described will stand as follows:

$$\begin{array}{r|l} X & X_{,} \\ X_{,}Q_{,} & \overline{Q_{,}} \\ \hline X-X_{,}Q_{,}=-X_{,,}; \end{array} \quad \begin{array}{r|l} X_{,} & X_{,,} \\ X_{,,}Q_{,,} & \overline{Q_{,,}} \\ \hline X_{,}-X_{,,}Q_{,,}=-X_{,,,}; \end{array} \quad \begin{array}{r|l} X_{,,} & X_{,,,} \\ X_{,,,}Q_{,,,} & \overline{Q_{,,,}} \\ \hline X_{,,}-X_{,,,}Q_{,,,}=-X_{,,,,}. \end{array}$$

We thus obtain the series of quantities

$$X, X_{,}, X_{,,}, X_{,,,}, X_{,,,,}, \ldots\ldots X_{m},$$

each of which is of a lower degree with respect to x than the preceding, and the last is altogether independent of x, that is, does not contain x.

We now substitute for x in the above functions any two numbers p and q, of which p is less than q. The substitution of p will give results either positive or negative. If we only take account of the *signs* of the results, we shall obtain a certain number of *variations* and a certain number of *permanences.*

The substitution of q for x will give a second series of signs, presenting a certain number of variations and permanences. The following, then, is

THE THEOREM OF STURM.

The difference between the number of variations of the first row of signs and that of the second, is equal to the number of real roots of the given equation comprised between p *and* q.

(321.) In order to simplify the demonstration of this theorem, we shall premise three Lemmas; and, for convenience, we shall call X the *primitive* function, and $X_{,}$, $X_{,,}$, $X_{,,,}$, &c., *auxiliary* functions.

LEMMA I. *If we substitute any number for* x *in the series of functions* X, $X_{,}$, $X_{,,}$, &c., *two consecutive functions can not both reduce to zero at the same time.*

For, from the method in which $X_{,}$, $X_{,,}$, &c., are obtained, we have the following equations:

$$X = X_{,}Q_{,} - X_{,,} \quad (1).$$
$$X_{,} = X_{,,}Q_{,,} - X_{,,,} \quad (2).$$
$$X_{,,} = X_{,,,}Q_{,,,} - X_{,,,,} \quad (3).$$
" " "
$$X_{m-2} = X_{m-1}Q_{m-1} - X_{m} \quad (m-1).$$

Now, if possible, suppose $X_{,}=0$, and $X_{,,}=0$; then, by equation (2), we shall have $X_{,,,}=0$. Also, since $X_{,,}=0$, and $X_{,,,}=0$; therefore, by equation (3), we must have $X_{,,,,}=0$; and, proceeding in the same manner, we shall find that $X_{m}=0$, which is *absurd*, since it was shown, *Art.* 320, that this final remainder must be independent of x, and must therefore remain unchanged for every value of x.

LEMMA II. *When one of the auxiliary functions vanishes for a particular value of* x, *the two adjacent functions must have contrary signs.*

For, by equation (3), we have

$$X_{,,} = X_{,,,}Q_{,,,} - X_{,,,,};$$

and if $X_{,,,}$ reduces to zero, then $X_{,,} = -X_{,,,,}$; that is, $X_{,,}$ and $X_{,,,,}$ have contrary signs.

LEMMA III. *If* a *is a root of the equation* $X=0$, *the signs of* X *and* $X_{,}$ *will constitute a variation for a value of* x *which is a little less than* a, *and a permanence for a value of* x *which is a little greater than* a.

For if we substitute $a+r$ for x in the equation $X=0$, the development of the function X, according to *Art.* 318, will be of the form

$$A + A'r + \text{other terms involving higher powers of } r.$$

Now if a is a root of the equation $X=0$, the first term of the development becomes zero, and there remains

$$A'r + \text{other terms involving higher powers of } r.$$

Also, if we substitute $a+r$ for x in the first derived polynomial, the development will contain

$$A' + \text{other terms involving}$$

Now we may take r so small that each of these developments shall have the same sign as its first term,

$$A'r \text{ and } A'.$$

Hence they must both have the *same* sign when r is positive, and *contrary* signs when r is negative. That is, the signs of the two functions X and $X_{,}$

constitute a variation for $x=a-r$,
and a permanence for $x=a+r$.

DEMONSTRATION OF THE THEOREM.

(322.) Suppose all the real roots of the equations

$$X=0,\ X_{,}=0,\ X_{,,}=0,\ X_{,,,}=0, \text{ \&c.},$$

to be arranged in a series in the order of magnitude, beginning with the least. Let p be less than the least of these roots, and let it increase continually until it becomes equal to q, which we suppose to be greater than the greatest of these roots. Now so long as p is *less* than any of the roots, no change of signs will occur from the substitution of p for x in any of these functions, *Art.* 315; but when p arrives at a root of any of the auxiliary equations, its substitution for x reduces that polynomial to zero, and neither the preceding nor succeeding function can vanish for the same value of x (Lemma I.), and these two adjacent functions have contrary signs (Lemma II.). Hence the entire number of variations of sign is not affected by the vanishing of any of the auxiliary functions; for the three adjacent functions must reduce to

$$+,\ 0,\ -, \text{ or } -,\ 0,\ +.$$

Here is one variation, and there will also be one variation if we supply the place of the 0 with either + or −; thus,

$$+,\ +,\ -, \text{ or } -,\ +,\ +,$$
$$+,\ -,\ -, \text{ or } -,\ -,\ +.$$

Suppose, now, p to pass from a number very little smaller, to a number very little greater than a root of the primitive equation

$$X=0,$$

the sign of X will be changed from + to −, or from − to +, *Art.* 315. The signs of X and $X_{,}$ constitute a variation before the change, and a permanence after the change (Lemma III.)

Hence the change of sign of the function X occasions a loss of one variation of sign.

Again, while p increases from a number very little smaller to a number very little greater than another root of the equation $X=0$, a second variation will be changed into a permanence, and so on for the other roots of the primitive equation.

Now, since all the real roots must be comprised within the limits $-\infty$ and $+\infty$, if we substitute these values for x in the series of functions X, $X_{,,}$ &c., the number of variations lost will indicate the whole number of real roots. A third supposition, that $x=0$, will show how many of these roots are positive and how many negative; and if we wish to determine smaller limits of the roots, we must try other numbers. It is generally best to make trial in the first instance of such numbers as are most convenient in computation, as, 1, 2, 10, &c.

EXAMPLES.

(323.) *Ex.* 1. How many real roots has the equation

$$x^3-6x^2+11x-6=0\,?$$

Here we have $\quad X_{,}=3x^2-12x+11.$

Dividing $x^3-6x^2+11x-6$ by $3x^2-12x+11$, as in the method for finding the greatest common divisor, *Art.* 251, we have for a remainder $-2x+4$. Hence, rejecting the factor [illegible] $X_{,,}=x-2$. Dividing $X_{,}$ by $X_{,,}$, we have for a remainder [illegible] Therefore, $X_{,,,}=+1$.

Hence we have

$$\begin{aligned} X\;\; &= x^3-\;6x^2+11x-6. \\ X_{,}\; &=3x^2-12x\;+11. \\ X_{,,} &= x\;-\;2. \\ X_{,,,} &=+1. \end{aligned}$$

If we substitute $-\infty$ for x in the first polynomial $x^3-6x^2+11x-6$, the sign of the result is $-$; substituting $-\infty$ for x in the second polynomial $3x^2-12x+11$, the sign of the result is $+$; substituting the same in $x-2$, the sign of the result is $-$; and $X_{,,,}$, being independent of x, will remain $+$ for every value of x, so that by supposing $x=-\infty$, we obtain the series of signs

$$-\;+\;-\;+.$$

Proceeding in the same manner for other assumed values of x, we shall obtain the following results:

Assumed Values of x.	Resulting Signs.	Variations.
$-\infty$	$- + - +$	giving 3 variations.
0	$- + - +$	" 3 "
$+.9$	$- + - +$	" 3 "
$+1$	$0 + - +$	" 2 "
$+1.1$	$+ + - +$	" 2 "
$+1.9$	$+ - - +$	" 2 "
$+2$	$0 - 0 +$	" 1 "
$+2.1$	$- - + +$	" 1 "
$+2.9$	$- + + +$	" 1 "
$+3$	$0 + + +$	" 0 "
$+3.1$	$+ + + +$	" 0 "
$+\infty$	$+ + + +$	" 0 "

Here the three roots of this equation are seen to be 1, 2, 3, and no change of sign in either function occurs by the substitution for x of any number less than 1; but when p exceeds 1, there is a change of sign in the original equation from $-$ to $+$, by which one variation is *lost.* When $p=2$, two of the functions disappear simultaneously, showing that 2 is a root of the second derived function as well as of the original equation, and a second variation of sign is *lost.* Also, when p becomes equal to 3, a third variation is lost; and there are no further changes of sign arising from the substitution of any numbers between 3 and $+\infty$.

There are *three* changes of sign of the primitive function, *two* of the first auxiliary function, and *one* of the second auxiliary function; but no variation is lost by the change of sign of any of the auxiliary functions; while every change of sign of the primitive function occasions a loss of one variation.

Ex. 2. How many real roots has the equation

$$x^3-5x^2+8x-1=0\,?$$

Here we find

$$X = x^3-5x^2+8x-1.$$
$$X_{\prime} = 3x^2-10x+8.$$
$$X_{\prime\prime} = 2x-31.$$
$$X_{\prime\prime\prime} = -2295.$$

When $x=-\infty$, the signs are $- + - -$, giving 2 variations;

$x=+\infty$, " $+ + + -$, " 1 "

Hence this equation has but one real root, and, consequent

ly, must have two imaginary roots. Moreover, it is easily proved that the real root lies between 0 and +1.

Ex. 3. How many real roots has the equation

$$x^4-2x^3-7x^2+10x+10=0\,?$$

Here we have

$$X \;=\; x^4-\;2x^3-\;7x^2+10x+10.$$
$$X_{\prime} \;=\; 4x^3-\;6x^2-14x+10;\ \text{or}\ 2x^3-3x^2-7x+5.$$
$$X_{\prime\prime} = 17x^2-23x-45$$
$$X_{\prime\prime\prime} = 152x-305.$$
$$X_{\prime\prime\prime\prime} = +524535.$$

When $x=-\infty$, the signs are $+-+-+$, giving 4 variations;
$x=+\infty$, " $+++++$, " 0 "

Hence the four roots of this equation are real.

If we try different values for x, we shall find that

When $x=-3$, the signs are $+-+-+$, giving 4 variations;
$x=-2$, " $-++-+$, " 3 "
$x=-1$, " $-++-+$, " 3 "
$x=\ 0$, " $++--+$, " 2 "
$x=+1$, " $+---+$, " 2 "
$x=+2$, " $+---+$, " 2 "
$x=+3$, " $+++++$, " 0 "

Hence this equation has one negative root between -2 and -3; one negative root between 0 and -1; and two positive roots between 2 and 3.

Ex. 4. How many real roots has the equation

$$x^3-7x+7=0\,?$$

Ans. Three: viz., two between 1 and 2, and one between -3 and -4.

Ex. 5. How many real roots has the equation

$$2x^4-20x+19=0\,?$$

Ans. Two.

Ex. 6. How many real roots has the equation

$$x^5+2x^4+3x^3+4x^2+5x-20=0\,?$$

Ans. One between 1 and 2.

Ex. 7. How many real roots has the equation

$$x^3+3x^2+5x-178=0\,?$$

Ans. One between 4 and 5.

Ex. 8. How many real roots has the equation

$$x^4-12x^2+12x-3=0\,?$$

Ans. Four.

Ex. 9. How many real roots has the equation

$$x^4-8x^3+14x^2+4x-8=0\,?$$

Ans. Four.

PROPOSITION XIV.

(324.) *To discover a method of elimination for equations of any degree.*

The principle of the greatest common divisor affords one of the most general methods for the elimination of unknown quantities from a system of equations.

Suppose we have two equations involving x and y reduced to the form of

$$A=0,$$
$$B=0.$$

If we proceed to find the greatest common divisor of A and B, we shall have, according to *Art.* 249,

$$A=QB+R.$$

But since A and B are each equal to zero, it follows that R must equal zero. Hence we see that, if we divide one of the polynomials by the other, as in the method of finding the greatest common divisor, each successive remainder may be put equal to zero. If we arrange the polynomials before division with reference to the letter x, we shall at last obtain a remainder which does not contain x; which remainder, being put equal to zero, is the equation from which x has been eliminated.

Ex. 1. Eliminate x from the equations

$$x^2+y^2-13=0,$$
$$x+y-5=0.$$

Divide the first polynomial by the second, as follows:

$$\begin{array}{l|l}
x^2+y^2-13 & x+y-5 \\
\underline{x^2+(y-5)x} & x-y+5 \\
\quad -(y-5)x+\ y^2-13 & \\
\quad \underline{-(y-5)x-\ y^2+10y-25} & \\
\qquad\qquad 2y^2-10y+12 = \text{remainder,} &
\end{array}$$

This remainder we have already proved must be equal to zero; that is,

$$2y^2-10y+12=0,$$

an equation from which x has been eliminated.

Ex. 2. Eliminate x from the equations

$$x^2+xy-56=0,$$
$$xy+2y^2-60=0.$$

Ans. $y^4-118y^2+1800=0.$

Ex. 3. Eliminate x from the equations

$$x^2+y^2-x-y-78=0,$$
$$xy+x+y-39=0.$$

Ans. $y^4+y^3-77y^2-273y+1404=0.$

Ex. 4. Eliminate x from the equations

$$x^2-3xy+y^2+y=0,$$
$$x^2-xy+1=0.$$

Ans. $y^4-5y^2+2y-1=0.$

If we have three equations containing three unknown quantities, we must first eliminate one of the unknown quantities by combining either of the equations with each of the others. We thus obtain two new equations involving but two unknown quantities, from which we may obtain a final equation involving but one unknown quantity.

Ex. 5. Eliminate x and y from the equations

$$xyz-c=0,$$
$$xz+xy+yz-b=0,$$
$$x+y+z-a=0.$$

Ans. $z^3-az^2+bz-c=0.$

Ex. 6. Eliminate x and y from the equations

$$x^2+y=7,$$
$$y^2+z=13,$$
$$z^2+x=18.$$

Ans. $z^8-72z^6+1930z^4-22824z^2+z+100476=0.$

SECTION XX.

SOLUTION OF NUMERICAL EQUATIONS.

(325.) We will first consider the method of finding the *integral* roots of an equation, and will begin with forming the equation whose roots are 2, 3, 4, and 5. This equation must be composed of the factors.

$$(x-2)\ (x-3)\ (x-4)\ (x-5)=0.$$

If we perform the multiplication (which is most expeditiously done by the method of detached coefficients shown in *Art.* 64), we obtain the equation

$$x^4-14x^3+71x^2-154x+120=0.$$

We know that this equation is divisible by $x-5$. Let us perform the division by the method of detached coefficients shown in *Art.* 80.

$$\begin{array}{l|l}
\text{A}\quad \text{B}\quad \text{C}\quad \text{D}\quad \text{V} & \quad a \\
1-14+71-154+120 & 1-5=\text{ divisor.} \\
\underline{1-\ \ 5} & \overline{1-9+26-24}=\text{ quotient.} \\
\quad -\ \ 9+71 & \\
\quad \underline{-\ \ 9+45} & \\
\qquad\quad +26-154 & \\
\qquad\quad \underline{+26-130} & \\
\qquad\qquad\quad -\ \ 24+120 & \\
\qquad\qquad\quad -\ \ 24+120. &
\end{array}$$

Supplying the powers of x, we obtain for a quotient

$$x^3-9x^2+26x-24=0.$$

This operation may be still further abridged, as follows:

Represent the root 5 by a, and the coefficients of the given equation by A, B, C, D, ... V.

We first multiply $-a$ by A, and subtract the product from B; the remainder, -9, we multiply by $-a$, and subtract the product from C; the remainder, $+26$, we multiply again by $-a$, and subtract from D; the remainder, -24, we multiply by $-a$, and, subtracting from V, nothing remains. If we take the root a with a positive sign, we may substitute addition for subtraction in the above statement; and if we set down only the successive remainders, the work will be as follows:

$$\begin{array}{l} \text{A}\quad \text{B}\quad \text{C}\quad \text{D}\quad \text{V}\ a \\ 1-14+71-154+120\underline{|5} \\ 1-\ 9+26-\ 24, \end{array}$$

and the rule will be,

Multiply A *by* a, *and add the product to* B; *set down the sum, multiply it by* a, *and add the product to* C; *set down the sum, multiply it by* a, *and add the product to* D, *and so on. The final product should be equal to the last term* V, *taken with a contrary sign.*

The coefficients above obtained are the coefficients of a cubic equation whose roots are 2, 3, 4. The equation may therefore be divided by $x-4$, and the operation will be as follows:

$$\begin{array}{l} 1-9+26-24|4 \\ 1-5+6. \end{array}$$

These, again, are the coefficients of a quadratic equation whose roots are 2 and 3. Dividing again by $x-3$, we have

$$\begin{array}{l} 1-5+6|3 \\ 1-2, \end{array}$$

which are the coefficients of the binomial factor $x-2$.

These three operations of division may be exhibited together as follows:

$$\begin{array}{l|l} 1-14+71-154+120 & 5, \text{ first divisor.} \\ 1-\ 9+26-\ 24 & 4, \text{ second divisor.} \\ 1-\ 5+\ 6 & 3, \text{ third divisor.} \\ 1-\ 2. & \end{array}$$

(326.) The method here explained will enable us to find all the integral roots of an equation. For this purpose, we make trial of different numbers in succession, all of which must be divisors of the last term of the equation. If any division leaves a remainder, we reject this divisor; if the division leaves no remainder, the divisor employed is a *root* of the equation. Thus, by a few trials, all the integral roots may be easily found.

Ex. 2. Find the seven roots of the equation

$$x^7+x^6-14x^5-14x^4+49x^3+49x^2-36x-36=0.$$

We take the coefficients separately, as in the last example, and try in succession all the divisors of 36, both positive and negative, rejecting such as leave a remainder. The operation is as follows:

1+1−14−14+49+49−36−36	1, first divisor.
1+2−12−26+23+72+36	2, second divisor.
1+4− 4−34−45−18	3, third divisor.
1+7+17+17+ 6	−1, fourth divisor.
1+6+11+ 6	−1, fifth divisor
1+5+ 6	−2, sixth divisor.
1+3	−3, seventh divisor.

Hence the seven roots are,

$$1,\ 2,\ 3,\ -1,\ -1,\ -2,\ -3.$$

Ex. 3. Find the six roots of the equation

$$x^6+5x^5-81x^4-85x^3+964x^2+780x-1584=0.$$

1+ 5−81− 85+964+ 780−1584	1.
1+ 6−75−160+804+1584	4.
1+10−35−300−396	6.
1+16+61+ 66	− 2.
1+14+33	− 3.
1+11	−11.

The six roots, therefore, are

$$1,\ 4,\ 6,\ -2,\ -3,\ -11.$$

Ex. 4. Find the five roots of the equation

$$x^5+6x^4-10x^3-112x^2-207x-110=0.$$

1+6−10−112−207−110	−1.
1+5−15− 97−110	−2.
1+3−21− 55	−5.
1−2−11	

Three of the roots, therefore, are

$$-1,\ -2,\ -5.$$

The two remaining roots may be found by the ordinary method of quadratic equations. Supplying the letters to the last coefficients, we have

$$x^2-2x-11=0.$$

Hence $$x=1\pm\sqrt{12}.$$

Ex. 5. Find the four roots of the equation

$$x^4+2x^3-7x^2-8x+12=0.$$

Ex. 6. Find the four roots of the equation

$$x^4-55x^2-30x+504=0.$$

Ex. 7. Find all the roots of the equation

$$x^4-25x^2+60x-36=0.$$

Ex. 8. Find all the roots of the equation

$$x^5+5x^4+x^3-16x^2-20x-16=0.$$

Ex. 9. Find all the roots of the equation

$$x^4-12x^3+47x^2-72x+36=0.$$

Ans. 1, 2, 3, and 6.

HORNER'S METHOD.

(327.) The preceding method furnishes the roots of an equation only when they are expressed by *whole numbers*. When the roots are incommensurable, we employ the following method, which is substantially the same as published by Horner in 1819.

The Theorem of Sturm, together with *Art.* 315, enables us to find the *integral part* of any real root of the equation proposed. We then transform the equation into another having its roots less than those of the preceding by the number just found, *Art.* 316. We discover again, by *Art.* 315, the first figure of the root of this equation, which will be the first decimal figure of the root of the original equation. Again, we transform the last equation into another having its roots less than those of the preceding by this decimal figure. We thus discover the second decimal figure of the root; and proceeding in this manner from one transformation to another, we are enabled to discover the successive figures of the root, and may carry the approximation to any degree of accuracy required.

Ex. 1. Find a root of the cubic equation

$$x^3+3x^2+5x=178.$$

We have found, page 278, that this equation has but one real root, and that it lies between 4 and 5. The first figure of the root, therefore, is 4. To ascertain the second figure, we transform the given equation into another in which the value of x is diminished by 4, which is done by substituting for x, $4+y$. We thus obtain

$$y^3+15y^2+77y=46.$$

The first figure of the root of this equation, according to *Art.* 315, is .5. Now transform the last equation into another in which the value of y is diminished by .5, which is done by substituting for y, $.5+z$. We thus obtain

$$z^3+16.5z^2+92.75z=3.625.$$

The first figure of the root of this equation is .03. We must now transform this equation into another in which the value of z is diminished by .03, which is done by substituting for z, $.03+v$. We thus obtain

$$v^3+16.59v^2+93.7427v=.827623.$$

The first figure of the root of this equation is .008.

In order to find the next figure, we must transform the last equation into another in which the value of v is diminished by .008, and so on.

(328.) This method would be very laborious if we were obliged to deduce the successive equations from each other by the ordinary method of substitution; but they may all be derived from each other by a very simple law. Thus, let

$$Ax^3+Bx^2+Cx=D \quad (1)$$

be any cubic equation; and let the first figure of its root be denoted by a, the second by a', the third by a'', and so on.

If we substitute a for x in equation (1), we shall have

$$Aa^3+Ba^2+Ca=D, \textit{ nearly}.$$

Whence $$a=\frac{D}{C+Ba+Aa^2} \quad (2).$$

If we put y for the sum of all the figures of the root except the first, we shall have $x=a+y$; and substituting this value for x in equation (1), we obtain

$$\left.\begin{array}{l}Aa^3+3Aa^2y+3Aay^2+Ay^3\\ \quad+\ Ba^2\ +2Bay\ +By^2\\ \qquad\quad+\ Ca\ \ +Cy\end{array}\right\}=D;$$

or, arranging according to the powers of y, we have

$$Ay^3+(B+3Aa)y^2+(C+2Ba+3Aa^2)y=D-Ca-Ba^2-Aa^3 \quad (3)$$

Let us put B' for the coefficient of y^2, C' for the coefficient of y, and D' for the right member of the equation, and we have

$$Ay^3+B'y^2+C'y=D' \quad (4).$$

This equation is of the same form as equation (1); and, proceeding in the same manner, we shall find

$$a'=\frac{D'}{C'+B'a'+Aa'^2} \quad (5),$$

where a' is the first figure of the root of equation (4), or the second figure of the root of equation (1).

Putting z for the sum of all the remaining figures, we have $y=a'+z$; and substituting this value in equation (4), we shall obtain a new equation of the same form, which may be written

$$Az^3+B''z^2+C''z=D'' \quad (6);$$

and in the same manner we might proceed with the remaining figures.

Equation (2) furnishes the value of the first figure of the root; equation (5) the second figure, and similar equations would furnish the remaining figures. Each of these expressions involves the unknown quantity which is sought, and might therefore appear to be useless in practice. When, however, the root has been already found to several decimal places, the value of the terms Ba and Aa^2 will be very small compared with C, and a will be very nearly equal to $\frac{D}{C}$. We may therefore employ C as an *approximate divisor*, which will probably furnish a new figure of the root. Thus, in the above example, all the figures of the root after the first are found by division

$$46 \div 77 \quad =.5.$$

$$3.625\div 92.75=.03.$$

$$.827\div 93.74=.008.$$

If we multiply the first coefficient A by a, the first figure of the root, and add the product to the second coefficient, we shall have

$$B+Aa \quad (7).$$

If we multiply this expression by a, and add the product to the third coefficient, we shall have

$$C+Ba+Aa^2 \quad (8).$$

If we multiply this expression by a, and subtract the product from D, we shall have

$$D-Ca-Ba^2-Aa^3,$$

which is the quantity represented by D′ *in equation* (4).

Again, multiplying the first coefficient by a, and adding the product to expression (7), we obtain

$$B+2Aa \quad (9).$$

Multiplying this expression by a, and adding the product to expression (8), we have

$$C+2Ba+3Aa^2,$$

which is the coefficient of y *in equation* (4).

Again, multiplying the first coefficient by a, and adding the product to expression (9), we have

$$B+3Aa,$$

which is the coefficient of y^2 *in equation* (4).

We have thus obtained the coefficients of the first transformed equation; and by operating in the same manner upon these coefficients, we shall obtain the coefficients of the second transformed equation, and so on; and the successive figures of the root are found by dividing D by C, D′ by C′, D″ by C″, and so on.

(329.) The preceding method is summed up in the following

RULE.

Represent the coefficients of the different terms by A, B, C, *and the right-hand member of the equation by* D. *Having found* a, *the first figure of the root, multiply* A *by* a, *and add the product to* B. *Set down the sum; multiply this sum by* a, *and add the product to* C. *Set down the sum; multiply it by* a, *and subtract the product from* D; *the remainder will be the* FIRST DIVIDEND.

Again, multiply A *by* a, *and add the product to the last number under* B. *Multiply this sum by* a, *and add the product to the last number under* C; *this last sum will be the* FIRST DIVISOR.

Again, multiply A *by* a, *and add the product to the last number under* B.

Find the second figure of the root by dividing the first dividend by the first divisor, and proceed with this second figure precisely as was done with the first figure.

The second figure of the root obtained by division will frequently furnish a result too large to be subtracted from the remainder D′, in which case we must assume a different figure. After the second figure of the root has been obtained, there will seldom be any further uncertainty of this kind.

The operation for finding a root of the equation

$$x^3+3x^2+5x=178,$$

will then proceed as follows:

A	B	C	D	a
1	+3	+5	=178	(4.5388=x.
	4	28	132	
	7	33	46 = 1st dividend.	
	4	44	42.375	
	11	77 = 1st divisor.	3.625 = 2d dividend.	
	4	7.75	2.797377	
	15.5	84.75	.827623 = 3d dividend.	
	.5	8.00	.751003872	
	16.0	92.75 = 2d divisor.	.076619128 = 4th dividend.	
	.5	.4959		
	16.53	93.2459		
	3	.4968		
	16.56	93.7427 = 3d divisor.		
	3	.132784		
	16.598	93.875484		
	8	.132848		
	16.606	94.008332 = 4th divisor.		

Having found one root, we may depress the equation

$$x^3+3x^2+5x-178=0$$

to a quadratic, by dividing it by $x-4.5388$. We thus obtain

$$x^2+7.5388x+39.2173=0,$$

where x is evidently imaginary, because q is negative and greater than $\frac{p^2}{4}$. See *Art.* 195.

After thus obtaining the root to five or six decimal places, *several more* figures will be correctly obtained by simply dividing the last dividend by the last divisor.

Ex. 2. Find all the roots of the equation

$$x^3+11x^2-102x=-181.$$

The first figure of one of the roots we readily find to be 3. We then proceed, according to the Rule, to obtain the root to four decimal places, after which two more will be obtained correctly by division.

A	B	C	D	a
1	+11	−102	=−181	(3.21312=x.
	3	42	−180	
	14	−60	−1 = 1st dividend.	
	3	51	−.992	
	17	−9 = 1st divisor.	−.008 = 2d dividend.	
	3	4.04	−.006739	
	20.2	−4.96	−.001261 = 3d dividend.	
	2	4.08	−.001217403	
	20.4	−0.88 = 2d divisor.	−.000043597 = 4th dividend.	
	2	.2061		
	20.61	−.6739		
	1	.2062		
	20.62	−.4677 = 3d divisor.		
	1	.061899		
	20.633	−.405801		
	3	.061908		
	20.636	−.343893 = 4th divisor.		

The two remaining roots may be found in the same way, or by depressing the original equation to a quadratic. Those roots are,

$$3.22952$$
$$-17.44265.$$

When a power of x is wanting in the proposed equation, we must supply its place with a cipher.

Ex. 3. Find all the roots of the cubic equation

$$x^3-7x=-7.$$

The work of the following example is exhibited in an abbreviated form. Thus, when we multiply A by a, and add the product to B, we set down simply this *result*. We do the same in the next column, thus dispensing with half the number of lines employed in the preceding example. Moreover, we may omit the ciphers on the left of the successive dividends, if we pay proper attention to the local value of the figures. Thus, it will be seen that in the operation for finding each successive figure of the root, the decimals under B increase *one* place, those under C increase *two* places, and those under D increase *three* places.

1 +0	−7	=−7 (1.356895867=x.
1	−6	−6
2	−4=1st div'r.	−1= 1st dividend.
3.3	−3.01	−.903
3.6	−1.93=2d div'r.	−97= 2d dividend.
3.95	−1.7325	86625
4.00	−1.5325= 3d div'r.	10375= 3d dividend.
4.056	−1.508164	9048984
4.062	−1.483792=4th div'r.	1326016= 4th dividend
4.0688	−1.48053696	1184429568
4.0696	−1.47728128=5th div'r.	141586432= 5th div'd.
4.07049	−1.4769149359	132922344231
4.07058	−1.4765485837= 6th div'r.	8664087769=6th div'd.

Having proceeded thus far, four more figures of the root, 5867, are found by dividing the sixth dividend by the sixth divisor.

We may find the two remaining roots by the same process; or, after having obtained one root, we may depress the equation

$$x^3-7x+7=0$$

to a quadratic equation, by dividing by $x-1.356895867$, and we shall obtain

$$x^2+1.356895867x-5.158833606=0.$$

Solving this equation, we obtain

$$x=-.678447933\pm\sqrt{5.619125204}.$$

Hence the three roots are . . . $\begin{cases} -3.048917, \\ 1.356896, \\ 1.692021. \end{cases}$

Ex. 4. Find a root of the equation

$$2x^3+3x^2=850.$$

2 3	0	=850	(7.0502562208
17	119	833	
31	336=1st divisor.	17= 1st dividend.	
45.10	338.2550	16.912750	
45.20	340.5150= 2d divisor.	87250= 2d dividend.	
45.3004	340.52406008	68104812016	
45.3008	340.53312024= 3d div.	19145187984= 3d div'd	
45.30130	340.5353853050	17026769265250	
45.30140	340.5376503750=4th div.	2118418718750= 4th div.	

Dividing the fourth dividend by the fourth divisor, we obtain the figures 62208, which make the root correct to the tenth decimal place.

The two remaining values of x may be easily shown to be imaginary.

When a negative root is to be found, we change the signs of the alternate terms of the equation, *Art.* 312, and proceed as for a positive root.

Ex. 5. Find a root of the equation

$$5x^3-6x^2+3x=-85.$$

Changing the signs of the alternate terms, it becomes

$$5x^3+6x^2+3x=+85.$$

5 +6	+3	+85	(2.16139.
16	35	70	
26	87= 1st divisor.	15= 1st dividend.	
36.5	90.65	9.065	
37.0	94.35= 2d divisor.	5.935= 2d dividend.	
37.80	96.6180	5.797080	
38.10	98.9040= 3d divisor.	137920= 3d dividend	
38.405	98.942405	98942405	
38.410	98.980815= 4th divisor.	38977595= 4th dividend	
38.4165	98.99233995	29697701985	
38.4180	99.00386535= 5th div'r.	9279893015=5th div'd.	

Hence one root of the equation

$$5x^3-6x^2+3x=-85$$

is -2.16139.

The same method is applicable to the extraction of the cube root of numbers.

Ex. 6. Let it be required to extract the cube root of 9; in other words, it is required to find a root of the equation

$$x^3=9.$$

1 0	0	9 (2.0800838.
2	4	8
4	12= 1st divisor.	1= 1st dividend.
6.08	12.4864	.998912
6.16	12.9792= 2d divisor.	1088= 2d dividend.
6.24008	12.9796992064	1038375936512
6.24016	12.9801984192= 3d d.	49624063488= 3d div.
6.240243	12.980217139929	38940651419787
6.240246	12.980235860667= 4th d.	10683412068213= 4th d

Ex. 7. Find all the roots of the equation

$$x^3-15x^2+63x-50=0.$$

Ans. $\begin{cases} 1.02804. \\ 6.57653. \\ 7.39543. \end{cases}$

Ex. 8. Find all the roots of the equation

$$x^3+9x^2+24x+17=0.$$

Ans. $\begin{cases} -1.12061. \\ -3.34730. \\ -4.53209. \end{cases}$

Ex. 9. Extract the cube root of 48228544.

Ans. 364.

Ex. 10. There are two numbers whose difference is 2, and whose product, multiplied by their sum, makes 120. What are those numbers?

Ex. 11. Find two numbers whose difference is 6, and such that their sum, multiplied by the difference of their cubes, may produce 5040.

Ex. 12. There are two numbers whose difference is 4; and

the product of this difference, by the sum of their cubes, is 3416. What are the numbers?

Ex. 13. Several persons form a partnership, and establish a certain capital, to which each contributes ten times as many dollars as there are persons in company. They gain 6 *plus* the number of partners per cent., and the whole profit is $392. How many partners were there?

Ex. 14. There is a number consisting of three digits such that the sum of the first and second is 9; the sum of the first and third is 12; and if the product of the three digits be increased by 38 times the first digit, the sum will be 336. Required the number.

Ans. { 636, or 725, or 814.

Ex. 15. A company of merchants have a common stock of $4775, and each contributes to it twenty-five times as many dollars as there are partners, with which they gain as much per cent. as there are partners. Now, on dividing the profit, it is found, after each has received six times as many dollars as there are persons in the company, that there still remains $126. Required the number of merchants.

Ans. 7, 8, or 9.

EQUATIONS OF THE FOURTH AND HIGHER DEGREES.

(330.) The method already explained for cubic equations is applicable to equations of every degree. For the fourth degree, we shall have one more column of products, but the operations are all conducted in the same manner, as will be seen from the following example.

Ex. 1. Find the four roots of the equation

$$x^4-8x^3+14x^2+4x=8.$$

By Sturm's Theorem, we have found that these roots are all real; three positive, and one negative.

We then proceed as follows:

1 −8	+14	+ 4	=8	(5.2360679.
−3	− 1	− 1	−5	
+2	+ 9	+44= 1st div'r.	13= 1st dividend.	
7	44	53.288	10.6576	
12.2	46.44	63.072= 2d div.	2.3424= 2d dividend.	
12.4	48.92	64.626747	1.93880241	
12.6	51.44	66.193068=3d d.	.40359759=3d dividend.	
12.83	51.8249	66.509117736	.399054706416	
12.86	52.2107	66.825633024= 4th d.	4542883584= 4th div.	
12.89	52.5974			
12.926	52.674956			
12.932	52.752548			

and by division we obtain the four figures 0679.

The other three roots may be found in the same manner

Hence the four roots are $\begin{cases} -\ .7320508, \\ \ \ .7639320, \\ 2.7320508, \\ 5.2360679. \end{cases}$

Ex. 2. Find a root of the equation

$$x^5+2x^4+3x^3+4x^2+5x=20.$$

We have found, by Sturm's Theorem, that this equation has a real root between 1 and 2.

We then proceed as follows:

1	+2		+3		+4		+5		+20		(1.125789
	3		6		10		15		15		
	4		10		20		35=1st divisor.		5= 1st dividend.		
	5		15		35		38.7171		3.87171		
	6		21		37.171		42.6585= 2d divisor.		1.12829=2d dividend.		
	7.1		21.71		39.414		43.5027	2016	.87005	44032	
	7.2		22.43		41.730		44.3566	2080= 3d divisor	.25823	55968= 3d dividend.	
	7.3		23.16		42.211	008	44.5731	44750625	.22286	5723753125	
	7.4		23.90		42.695	032	44.7902	83203125= 4th divisor.	3536	9873046875= 4th dividend.	
	7.5	2	24.05	04	43.182	080					
	7.5	4	24.20	12	43.304	790125					
	7.5	6	24.35	24	43.427	690500					
	7.5	8	24.50	40							
	7.6	05	24.54	2025							
	7.6	10	24.58	0075							

Dividing the fourth dividend by the fourth divisor, we obtain the figures 789.

When we wish to obtain a root correct to a limited number of places, we may save much of the labor of the operation by cutting off all figures beyond a certain decimal. Thus if, in the example above, we cut off all beyond five decimal places in the successive dividends, and all beyond four decimal places in the divisors, it will not affect the first six decimal places in the root.

Ex. 3. Find the roots of the equation

$$x^4-12x^2+12x=3.$$

$$Ans. \begin{cases} -3.907378, \\ +\ .443277, \\ +\ .606018, \\ +2.858083. \end{cases}$$

Ex. 4. Find the roots of the equation

$$x^4-16x^3+79x^2-140x=-58.$$

$$Ans. \begin{cases} +0.58579, \\ +3.35425, \\ +3.41421, \\ +8.64575. \end{cases}$$

Ex. 5. Find the roots of the equation

$$x^5-20x^4+150x^3-520x^2+806x=407.$$

$$Ans. \begin{cases} +0.934685, \\ +3.308424, \\ +3.824325, \\ +4.879508, \\ +7.053058. \end{cases}$$

Ex. 6. Required the fourth root of 18339659776.

Ans. 368.

Ex. 7. Required the fifth root of 26286674882643.

Ans. 483.

Ex. 8. There is a number consisting of four digits such that the sum of the first and second is 9; the sum of the first and third is 10; the sum of the first and fourth is 11; and if the product of the four digits be increased by 36 times the product of the first and third, the sum will be equal to 3024 diminished by 300 times the first digit. Required the number.

$$Ans. \begin{cases} 6345, \\ \text{or } 7234, \\ \text{or } 8123, \\ \text{or } 9012. \end{cases}$$

RESOLUTION OF EQUATIONS BY APPROXIMATION.

(331.) The method of Horner for finding the incommensurable roots of a numerical equation is generally better than any other; nevertheless, the method by approximation may sometimes be preferred. We shall explain the method of Newton, and that of Double Position.

METHOD OF NEWTON.

This method supposes that we have already determined nearly the value of one root; that we know, for example, that such a value exceeds a, and that it is less than $a+1$. In this case, if we suppose the exact value $=a+y$, we are certain that y expresses a proper fraction. Now, as y is less than unity, the square of y, its cube, and, in general, all its higher powers, will be much less with respect to unity; and for this reason, since we only require an approximation, they may be neglect ed in the calculation. When we have nearly determined the fraction y, we shall know more exactly the root $a+v$; from which we proceed to determine a new value still more exact, and we may continue the approximation as far as we please.

We will illustrate this method by an easy example, requiring by approximation the root of the equation

$$x^2=20.$$

Here we perceive that x is greater than 4, and less than 5. If we suppose $x=4+y$, we shall have

$$x^2=16+8y+y^2=20.$$

But, as y^2 must be quite small, we shall neglect it, and we have

$$16+8y=20, \text{ or } 8y=4.$$

Whence $y=.5$, and $x=4.5$, which already approaches near the true root. If we now suppose $x=4.5+z$, we are sure that

z expresses a fraction much smaller than y, and that we may neglect z^2 with greater propriety. We have, therefore,

$$x^2=20.25+9z=20, \text{ or } 9z=-.25.$$

Consequently $z=-.0278.$

Therefore, $x=4.5-.0278=4.4722.$

If we wish to approximate still nearer to the true value, we must make $x=4.4722+v$, and we should have

$$x^2=20.00057284+8.9444v=20.$$

So that $8.9444v=-.00057284.$

Whence $v=-.0000640.$

Therefore, $x=4.4722-.0000640=4.4721360,$

a value which is correct to the last decimal place.

(332.) The preceding method is expressed in the following

RULE.

Find by trial a number (a) *nearly equal to the root sought, and represent the true root by* a+y.

Substitute a+y *for* x *in the given equation, and there will result a new equation containing only* y *and known quantities.*

Reject all the terms of this equation which contain the second or higher powers of y, *and the approximate value of* y *will then be given by a simple equation.*

Having applied this correction to the assumed root, the operation must be repeated with the corrected value of a, when a second correction will be obtained which will give a nearer value of the root, and the process may be repeated as often as is thought necessary.

EXAMPLES.

Ex. 1. Find a root of the equation

$$x^3+2x^2+3x=50.$$

If we substitute $a+y$ for x in this equation, and reject all the terms containing the higher powers of y, we shall have

$$a^3+3a^2y+2a^2+4ay+3a+3y=50.$$

Whence $$y=\frac{50-a^3-2a^2-3a}{3a^2+4a+3}.$$

We find by trial that x is nearly equal to 3. If we substitute 3 for a, we shall have

$$y=-\frac{2}{21}.$$

Whence $x=2.9$ nearly.

And if we substitute this new value instead of a, we shall find another still more exact.

Ex. 2. Find a root of the equation

$$x^5-6x=10.$$

If we make $x=a+y$, we shall have

$$a^5+5a^4y-6a-6y=10.$$

Therefore, $$y=\frac{10+6a-a^5}{5a^4-6.}$$

Assume $a=2$, and we obtain

$$y=-\frac{5}{37}, \text{ or } -0.14.$$

Hence $x=1.86$ nearly.

If we assume $a=1.86$, we have

$$y=\frac{10+11.16-22.262}{59.844-6}=-.021.$$

Hence $x=1.839$ nearly.

If we assume $a=1.839$, we shall have

$$y=\frac{10+11.034-21.033352}{57.18694-6}=.00001266.$$

Therefore, $x=1.83901266$.

Ex. 3. Given $x^3-9x=10$, to find one value of x by approximation.

Ans. $x=3.4494897$.

Ex. 4. Given $x^3+9x^2+4x=80$, to find one value of x by approximation.

Ans. $x=2.4721359$

METHOD OF DOUBLE POSITION.

(333.) Another method of finding the roots of an equation is by the rule of Double Position.

Substitute in the given equation two numbers as near the true root as possible, and observe the separate results. Then state the following proportion:

As the difference of these results,

Is to the difference of the two assumed numbers,

So is the error of either result,

To the correction required in the corresponding assumed number.

This being added to the number when too small, or subtracted from it when too great, will give the true root *nearly.* The number thus found, combined with any other that may be supposed to approach still nearer to the true root, may be assumed for another operation, which may be repeated till the root is determined to any degree of accuracy required.

EXAMPLES.

Ex. 1. Given $x^3+x^2+x=100$, to find an approximate value of x.

Having ascertained, by trial, that x is more than 4, and less than 5, we substitute these two numbers in the given equation, and calculate the results.

By the first supposition,	$x=4$, $x^2=16$, $x^3=64$	By the second supposition,	$x=5$, $x^2=25$, $x^3=125$
Result,	84	Result,	155.

Then 155−84 : 5−4 : : 100−84 : .22.

Therefore, 4+.22, or 4.22, approximates nearly to the true root.

If, now, 4.2 and 4.3 be taken as the assumed numbers, and substituted in the given equation, we shall obtain the value of $x=4.264$ nearly.

Again, assuming 4.264 and 4.265, and proceeding in the same manner, we shall find $x=4.2644299$ very nearly.

This rule is founded on the supposition that the differences in the results are proportioned to the differences in the assumed numbers. This supposition is not strictly correct, but if we employ numbers near the true values, the error is generally not very great, and it becomes less and less the further we carry the approximation.

Ex. 2. Given $x^3+2x^2-23x=70$, to find one value of x.

Ans. $x=5.13458$.

Ex. 3. Given $x^4-3x^2-75x=10000$, to find one value of x.

Ans. $x=10.2610$.

Ex. 4. Given $x^6+3x^4+2x^3-3x^2-2x=2$, to find one value of x.

Ans. $x=1.059109$.

(334.) We will conclude this Section by finding some of *the different roots of unity.*

Ex. 1. Find the two roots of the equation $x^2=1$, or the square roots of unity.

Extracting the square root, we find

$$x=+1, \text{ or } -1.$$

Ex. 2. Find the three roots of the equation $x^3=1$, or the cube roots of unity.

Since one root of this equation is $x=1$, the equation $x^3-1=0$ must be divisible by $x-1$; and dividing, we obtain

$$x^2+x+1=0;$$

whence $$x=-\tfrac{1}{2}\pm\tfrac{1}{2}\sqrt{-3}, \text{ or } \frac{-1\pm\sqrt{-3}}{2}.$$

Hence the required roots are

$$+1, \frac{-1+\sqrt{-3}}{2}, \frac{-1-\sqrt{-3}}{2},$$

which are the cube roots of unity.

These results may be easily verified. We have seen, on page 259, that the cube of $-1\pm\sqrt{-3}$ is 8, which, divided by 8, the cube of the denominator, gives $+1$, as required.

Ex. 3. Find the four roots of the equation $x^4=1$, or the fourth roots of unity.

The square root of this equation is

$$x^2=+1, \text{ or } =-1.$$

Hence the required roots are

$$+1, \ -1, \ +\sqrt{-1}, \ -\sqrt{-1}.$$

Ex. 4. Find the five roots of the equation $x^5=1$.

Since one root of this equation is $x=1$, the equation x^5-1 must be divisible by $x-1$; and dividing, we obtain

$$x^4+x^3+x^2+x+1=0.$$

Dividing, again, by x^2, we have

$$x^2+x+1+\frac{1}{x}+\frac{1}{x^2}=0 \quad (1).$$

Now put $$z=x+\frac{1}{x}.$$

Whence $$z^2=x^2+2+\frac{1}{x^2},$$

which, being substituted in equation (1), gives

$$z^2+z-1=0.$$

This equation, solved by the usual method, gives

$$z=-\tfrac{1}{2}+\tfrac{1}{2}\sqrt{5}, \text{ or } z=-\tfrac{1}{2}-\tfrac{1}{2}\sqrt{5}.$$

The values of x, deduced from the equation

$$z=x+\frac{1}{x},$$

or $$x^2-zx=-1,$$

are

$$x=\frac{z}{2}+\sqrt{\frac{z^2-4}{4}}, \text{ and } x=\frac{z}{2}-\sqrt{\frac{z^2-4}{4}},$$

from which, by substituting the value of z, we obtain

$$x=\tfrac{1}{4}[\sqrt{5}-1\pm\sqrt{-10-2\sqrt{5}}],$$

or $$=-\tfrac{1}{4}[\sqrt{5}+1\mp\sqrt{-10+2\sqrt{5}}].$$

Hence the five fifth roots of unity are

$$1$$

$$\tfrac{1}{4}[\sqrt{5}-1+\sqrt{-10-2\sqrt{5}}].$$

$$\tfrac{1}{4}[\sqrt{5}-1-\sqrt{-10-2\sqrt{5}}].$$

$$-\tfrac{1}{4}[\sqrt{5}+1-\sqrt{-10+2\sqrt{5}}].$$

$$-\tfrac{1}{4}[\sqrt{5}+1+\sqrt{-10+2\sqrt{5}}].$$

Ex. 5. Find the six roots of the equation $x^6=1$.

These are found by taking the square roots of the cube roots Hence we have,

$$+1, -1, \tfrac{1}{2}\pm\tfrac{1}{2}\sqrt{-3}, -\tfrac{1}{2}\pm\tfrac{1}{2}\sqrt{-3}.$$

Thus we see that unity has *two* square roots, *three* cube roots, *four* fourth roots, *five* fifth roots, *six* sixth roots, and, generally, the nth root of unity admits of n different algebraic values. As, however, most of these roots are imaginary, they can not be found by Horner's Method.

SECTION XXI.

LOGARITHMS.

(335.) In a system of logarithms, all numbers are considered as the powers of some one number, arbitrarily assumed, which is called the *base* of the system; and *the exponent of that power of the base which is equal to any given number is called the logarithm of that number.*

Thus, if a be the base of a system of logarithms, and $a^2=N$, then 2 is the logarithm of N; that is, 2 is the exponent of the power to which the base (a) must be raised to equal N.

If $a^3=N'$, then 3 is the logarithm of N' for the same reason; and if $a^x=N''$, then x is called the logarithm of N'' in the system whose base is a.

The base of the common system of logarithms (called, from their inventor, Briggs' Logarithms) is the number 10. Hence in this system all numbers are to be regarded as powers of 10. Thus, since

$10^0=1$,	0 is the logarithm of	1	in Briggs' system.
$10^1=10$,	1 "	10	"
$10^2=100$,	2 "	100	"
$10^3=1000$,	3 "	1000	"
$10^4=10000$,	4 "	10000	"
&c.,	&c.,		&c.

From this it appears that, in Briggs' system, the logarithm of every number between 1 and 10 is some number between 0 and 1, *i. e.*, is a proper fraction. The logarithm of every number between 10 and 100 is some number between 1 and 2, *i. e.*, is 1 plus a fraction. The logarithm of every number between

100 and 1000 is some number between 2 and 3, *i. e.*, is 2 plus a fraction, and so on.

(336.) The preceding principles may be extended to fractions by means of negative exponents. Thus,

10^{-1} or $\frac{1}{10}$ $=0.1$; therefore, -1 is the logarithm of .1 in Briggs' system.

10^{-2} or $\frac{1}{100}$ $=0.01$;	"	-2	"	.01
10^{-3} or $\frac{1}{1000}$ $=0.001$;	"	-3	"	.001
10^{-4} or $\frac{1}{10000}$ $=0.0001$;	"	-4	"	.0001.

Hence it appears that the logarithm of every number between 1 and .1 is some number between 0 and -1, or may be represented by -1 plus a fraction; the logarithm of every number between .1 and .01 is some number between -1 and -2, or may be represented by -2 plus a fraction; the logarithm of every number between .01 and .001 is some number between -2 and -3, or is equal to -3 plus a fraction, and so on.

(337.) The logarithms of most numbers, therefore, consist of an integer and a fraction. The integral part is called the *characteristic*, and may always be known from the following

RULE.

The characteristic of the logarithm of any number greater than unity, is one less than the number of integral figures in the given number.

Thus the logarithm of 297 is 2 plus a fraction; that is, the *characteristic* of the logarithm of 297 is 2, which is one less than the number of integral figures. The characteristic of the logarithm of 5673 is 3; of 73254 is 4, &c.

The characteristic of the logarithm of a decimal fraction is a negative number, and is equal to the number of places by which its first significant figure is removed from the place of units.

Thus the logarithm of .0046 is 3 plus a fraction; that is, the characteristic of the logarithm is -3, the first significant figure, 4, being removed three places from units.

In a series of fractions continually decreasing, the negative logarithms continually increase. Hence, if the fraction is infinitely small, its logarithm will be infinitely great; that is, in Briggs' system, *the logarithm of zero is infinite and negative.*

GENERAL PROPERTIES OF LOGARITHMS.

(338.) Let N and N′ be any two numbers, x and x' their respective logarithms, and a the base of the system. Then, by the definition, *Art.* 335,

$$N = a^x \quad (1).$$

Also

$$N' = a^{x'} \quad (2).$$

Multiplying together equations (1) and (2), we obtain

$$NN' = a^x a^{x'}$$
$$= a^{x+x'}.$$

Therefore, according to the definition of logarithms, $x+x'$ is the logarithm of NN′, since $x+x'$ is the exponent of that power of the base a which is equal to NN′; hence

PROPERTY I.

The logarithm of the product of two or more factors is equal to the sum of the logarithms of those factors.

Hence we see that if it is required to multiply two or more numbers by each other, we have only to add their logarithms; the sum will be the logarithm of their product. We must then look in the Table for the number answering to that logarithm, in order to obtain the required product.

EXAMPLES.

Ex. 1. Find the product of 8 and 9 by means of logarithms.

On page 318, the logarithm of 8 is given	0.903090
" 9 "	0.954243
The sum of these two logarithms is	1.857333,

which, according to the same Table, is seen to be the logarithm of 72.

Ex. 2. Find the continued product of 2, 5, and 14 by means of logarithms.

Ex. 3. Find the continued product of 1, 2, 3, 4, and 5 by means of logarithms.

(339.) If, instead of multiplying, we divide equation (1) by equation (2), we shall obtain

$$\frac{N}{N'}=\frac{a^x}{a^{x'}}=a^{x-x'}.$$

Therefore, according to the definition, $x-x'$ is the logarithm of $\frac{N}{N'}$, since $x-x'$ is the exponent of that power of the base a which is equal to $\frac{N}{N'}$; hence,

PROPERTY II.

The logarithm of a fraction, or of the quotient of one number divided by another, is equal to the logarithm of the numerator, minus the logarithm of the denominator.

Hence we see that if we wish to divide one number by another, we have only to subtract the logarithm of the divisor from that of the dividend; the difference will be the logarithm of their quotient.

EXAMPLES.

Ex. 1. It is required to divide 108 by 12 by means of logarithms.

The logarithm of 108 is	2.033424
" 12	1.079181
The difference is	0.954243,

which is the logarithm corresponding to the number 9.

Ex. 2. Divide 133 by 7 by means of logarithms.

Ex. 3. Divide 136 by 17 by means of logarithms.

Ex. 4. Divide 135 by 15 by means of logarithms.

The preceding examples are designed to illustrate the properties of logarithms. In order to exhibit fully their *utility* in computation, it would be necessary to employ larger numbers; but that would require a more extensive Table than the one given on page 318.

(340.) Logarithms are attended with still greater advantages in the involution of powers and in the extraction of roots. For if we raise both members of equation (1) to the mth power, we obtain

$$N^m=a^{mx}.$$

Therefore, according to the definition, mx is the logarithm of N^m, since mx is the exponent of that power of the base which is equal to N^m; hence

PROPERTY III.

The logarithm of any power of a number is equal to the logarithm of that number multiplied by the exponent of the power.

EXAMPLES.

Ex. 1. Find the third power of 4 by means of logarithms.

The logarithm of 4 is	0.602060
Multiply by	3
The product is	1.806180,

which is the logarithm of 64.

Ex. 2. Find the fourth power of 3 by means of logarithms.

Ex. 3. Find the seventh power of 2 by means of logarithms.

Ex. 4. Find the third power of 5 by means of logarithms.

(341.) Also, if we extract the mth root of both members of equation (1), we shall obtain

$$N^{\frac{1}{m}}=a^{\frac{x}{m}};$$

therefore, according to the definition, $\frac{x}{m}$ is the logarithm of $N^{\frac{1}{m}}$; hence

PROPERTY IV.

The logarithm of any root of a number is equal to the logarithm of that number divided by the index of the root.

EXAMPLES.

Ex. 1. Find the square root of 81 by means of logarithms

The logarithm of 81 is	1.908485
Divided by 2	
The quotient is	.954243,

which is the logarithm of 9.

Ex. 2. Find the square root of 121 by means of logarithms.

Ex. 3. Find the sixth root of 64 by means of logarithms.

Ex. 4. Find the third root of 125 by means of logarithms.

The preceding examples will suffice to show, that if we had tables which gave the logarithms of all numbers, they would prove highly useful when we have occasion to perform frequent multiplications, divisions, involutions, and extraction of roots.

(342.) The following examples will show the application of some of the preceding principles.

Ex. 1. $\log.\ (abcd) = \log.\ a + \log.\ b + \log.\ c + \log.\ d.$

Ex. 2. $\log.\ \left(\frac{abc}{de}\right) = \log.\ a + \log.\ b + \log.\ c - \log.\ d - \log.\ e.$

Ex. 3. $\log.\ (a^m b^n c^p) = m \log.\ a + n \log.\ b + p \log.\ c.$

Ex. 4. $\log.\ \left(\frac{a^m b^n}{c^p}\right) = m \log.\ a + n \log.\ b - p \log.\ c.$

Ex. 5. $\log.\ (a^2 - x^2) = \log.\ [(a+x)\ (a-x)] = \log.\ (a+x) + \log.\ (a-x).$

Ex. 6. $\log.\ \sqrt{a^2 - x^2} = \frac{1}{2} \log.\ (a+x) + \frac{1}{2} \log.\ (a-x).$

Ex. 7. $\log.\ (a^3 \sqrt[4]{a^3}) = \log.\ \left(a^{\frac{15}{4}}\right) = \frac{15}{4} \log.\ a.$

(343.) We shall presently explain a method by which logarithms may be computed. We may observe, however, that it is not necessary to compute the logarithms of all numbers independently. From the logarithms of a few numbers, we may readily derive the logarithms of a great many other numbers.

We have seen, in *Art.* 338, that the logarithm of a product is found by adding together the logarithms of the factors. Let us represent the logarithm of 2 by x; then, since the logarithm of 10 is 1, we shall have

$$\log.\ 20 \ = x+1, \qquad \log.\ 20000 \ =$$
$$\log.\ 2000 = x+3, \qquad \log.\ 2000000 =, \ \&c.$$

We have seen, in *Art.* 340, that the logarithm of any power of a number is equal to the logarithm of that number multiplied by the exponent of the power.

Hence, log. 4 $=2x$, log. 32 $=$
log. 16$=4x$, log. 128$=$, &c.

Hence we find, also, that

log. 40 $=2x+1$, log. 4000 $=$
log. 400 $=2x+2$, log. 40000 $=$, &c.
log. 80 $=3x+1$, log. 8000 $=$
log. 800 $=3x+2$, log. 80000 $=$, &c.
log. 160 $=4x+1$, log. 16000 $=$
log. 1600$=4x+2$, log. 160000$=$, &c.

We have seen, in *Art.* 339, that the logarithm of a fraction is equal to the logarithm of the numerator minus the logarithm of the denominator. Hence, log. 5$=$ log. $(\frac{10}{2})=1-x$.

Hence, log. 50 $=2-x$, log. 5000 $=$
log. 500 $=3-x$, log. 50000 $=$, &c.
log. 25 $=2-2x$, log. 625 $=$
log. 125 $=3-3x$, log. 3125 $=$, &c.
log. 250 $=3-2x$, log. 25000 $=$
log. 2500 $=4-2x$, log. 250000 $=$, &c.
log. 1250 $=4-3x$, log. 125000 $=$
log. 12500$=5-3x$, log. 1250000$=$, &c.
log. 6250 $=5-4x$, log. 625000 $=$
log. 62500$=6-4x$, log. 6250000$=$, &c.

(343.) So, also, from the logarithm of 3 we might easily derive a great number of other logarithms. From the Table on page 318, we find the logarithm of 3 to be .477121: it is required to derive from this the logarithm of 30.

Required the logarithm of 3000.

Required the logarithm of 9.

Required the logarithm of 27.

Required the logarithm of 81.

Required the logarithm of 90.

Required the logarithm of 270.

Required the logarithm of 900.

From the same Table, we find the logarithm of 2 to be .301030: it is required, by the aid of the logarithms of 3 and 2, to obtain the logarithm of 6.

Required the logarithm of 12.

Required the logarithm of 15.

Required the logarithm of 18.

From the same Table, we find the logarithm of 5 to be .698970. It is required from this to deduce the logarithm of 50.

Required the logarithm of 500.

Required the logarithm of 5000.

From the same Table, we find the logarithm of 95 to be 1.977724. The logarithm of 9.5, or $\frac{95}{10}$, is equal to the logarithm of 95 minus the logarithm of 10.

Hence the logarithm of 9.5 is 0.977724.

Also, the logarithm of 950 is 2.977724.

Hence the decimal part of the logarithm of any number is the same as that of the number multiplied or divided by 10, 100, 1000, &c.

Prime numbers are such as can not be decomposed into factors; as, 2, 3, 5, 7, 11, 13, 17, &c. All other numbers arise from the multiplication of prime numbers. If, therefore, we knew the logarithms of all the prime numbers, we could find the logarithms of all other numbers by simple addition.

(345.) We will now explain a method by which the logarithm of any number may be computed.

If a series of numbers be taken in *Geometrical progression*, their logarithms will form a series in *Arithmetical progression*. Thus, take the geometrical series

1, 10, 100, 1000, 10000, 100000,

their logarithms are

0, 1, 2, 3, 4, 5,

forming an arithmetical series.

If, now, we find a geometrical mean between any two numbers in the first series, its logarithm will be the arithmetical mean between the two corresponding numbers in the lower series.

Find, for example, a geometrical mean between 1 and 10. It will be the square root of 10, or 3.162277. The arithmetical mean between 0 and 1 is 0.5.

Therefore, the logarithm of 3.162277 is 0.5.

Find, again, a geometrical mean between 3.162277 and 10 which is 5.623413. Find, also, the arithmetical mean between 0.5 and 1, which is 0.75.

Therefore, the logarithm of 5.623413 is 0.75.

Find, now, a geometrical mean between 3.162277 and 5.623413, which is 4.216964. Its logarithm will be the arithmetical mean between 0.5 and 0.75, which is 0.625.

Therefore, the logarithm of 4.216964 is 0.625.

Find, again, a geometrical mean between 4.216964 and 5.623413, which is 4.869674. Its logarithm will be the arithmetical mean between 0.625 and 0.75, which is 0.6875.

Thus we have found the logarithms of four new numbers, and in this manner we might proceed to construct a table of logarithms. It will be observed that these numbers are all *fractional*, whereas it is most convenient to have the logarithms of *integers*. By pursuing this method, however, we might eventually find the logarithm of a whole number; as, for example, 5. For we have already found the logarithm of

5.623413 to be 0.75,

and the logarithm of 4.869674 " 0.6875.

One of these numbers is *greater* than 5, and the other *less*. A geometrical mean between them is 5.232991, which is too great; but the mean between this result and the last of the two preceding is 5.048065, which is already a close approximation. By pursuing the same method, we may come nearer and nearer to the number 5, until at last, after finding twenty-two geometrical means, the difference is inappreciable in the sixth decimal place, and we obtain

the logarithm of 5 equal to 0.69897;

and, by a like process, the logarithm of any other number may be found.

(346.) Hence, to compute the logarithm of any number, we have the following

RULE.

Take the geometrical series 1, 10, 100, 1000, 10000, &c., *and apply to it the arithmetical series* 0, 1, 2, 3, 4, &c., *as logarithms.*

Find a geometrical mean between 1 *and* 10, 10 *and* 100, *or any other two terms of the first series between which the proposed number lies.*

Between the mean thus found and the nearest term of the first series, find another geometrical mean in the same manner, and so on, till you approach as near as is necessary to the number whose logarithm is sought.

Find, also, as many arithmetical means between the corresponding terms 0, 1, 2, 3, 4, &c., *of the other series, in the same order as the geometrical ones were found; the last of these will be the logarithm answering to the number required.*

In this manner were the logarithms of all the prime numbers at first computed; but much more expeditious methods have since been devised.

Having obtained the logarithm of 5, it is easy to find the logarithm of 2. For the logarithm of $2=\log.\ (\frac{10}{5})=\log.\ 10-\log.\ 5=1-0.69897=0.30103$.

LOGARITHMIC SERIES.

(347.) The preceding method of computing logarithms is very laborious in practice. It is found much more convenient to express the logarithm of a number in the form of a *series.*

Let x be a number whose logarithm is required to be developed in a series, and let us employ the method of Unknown Coefficients. It is plain that we can not assume

$$\log.\ x=A+Bx+Cx^2+,\ \&c.\ ;$$

for when we make $x=0$, the first member reduces to infinity, while the second member reduces to A, a finite quantity Neither can we suppose

$$\log.\ x=Ax+Bx^2+Cx^3+,\ \&c.:$$

for when we make $x=0$, we have

log. 0 (which is infinite), equal to zero,

which is absurd.

But if we suppose

$$\log.\ (1+x)=Ax+Bx^2+Cx^3+Dx^4+,\ \&c. \quad (1),$$

when we make $x=0$, the equation becomes

log. 1, equal to zero

which is conformable to *Art.* 335.

Let us also assume

$$\log.\ (1+z)=Az+Bz^2+Cz^3+Dz^4+,\ \&c. \quad (2).$$

Subtracting equation (2) from (1), we obtain

$$\log.\ (1+x)-\log.\ (1+z)=A(x-z)+B(x^2-z^2)+C(x^3-z^3)+,\ \&c. \quad (3).$$

The second member of this equation is divisible by $x-z$, *Art.* 76; we will reduce the first member to a form in which it shall also be divisible.

We have $\log.\ (1+x)-\log.\ (1+z)=\log.\ \left(\frac{1+x}{1+z}\right)=$ $\log.\ \left(1+\frac{x-z}{1+z}\right)$.

Now, since $\frac{x-z}{1+z}$ may be regarded as a single quantity, v, we may develop $\log.\ (1+v)$ in the same manner as $\log.\ (1+x)$, which gives

$$\log.\ \left(1+\frac{x-z}{1+z}\right)=A.\frac{x-z}{1+z}+B\left(\frac{x-z}{1+z}\right)^2+C\left(\frac{x-z}{1+z}\right)^3+,\ \&c.$$

This last series must be identical with the one which we have already obtained for $\log.\ \left(1+\frac{x-z}{1+z}\right)$, or its equal, $\log.\ (1+x)-\log.\ (1+z)$, in equation (3); and since the terms of both are divisible by $x-z$, by canceling this common factor, we obtain

$$A.\frac{1}{1+z}+B\frac{x-z}{(1+z)^2}+C\frac{(x-z)^2}{(1+z)^3}+,\ \&c.,\ =A+B(x+z)+C(x^2+xz+z^2)+,\ \&c.$$

Since this equation, like the preceding, must be verified for all values of x and z, the equality must subsist when $x=z$. But on this hypothesis, all the terms of the first series vanish except one, and we have

$$\frac{A}{1+x}=A+2Bx+3Cx^2+4Dx^3+5Ex^4+,\ \&c.;$$

or, performing the division indicated in the first member, we obtain

$$A(1-x+x^2-x^3+x^4-\ldots)=A+2Bx+3Cx^2+4Dx^3+\ldots$$

Therefore, according to the principle of *Art.* 302, we have the equations

$$\mathrm{A}=\mathrm{A},$$

$$-\mathrm{A}=2\mathrm{B};\ \text{whence}\ \mathrm{B}=-\frac{\mathrm{A}}{2}.$$

$$\mathrm{A}=3\mathrm{C};\quad \text{"}\quad \mathrm{C}=+\frac{\mathrm{A}}{3}.$$

$$-\mathrm{A}=4\mathrm{D};\quad \text{"}\quad \mathrm{D}=-\frac{\mathrm{A}}{4}.$$

$$\mathrm{A}=5\mathrm{E};\quad \text{"}\quad \mathrm{E}=+\frac{\mathrm{A}}{5}.$$

The law of the series is obvious; and hence, substituting the values of B, C, D, &c., in equation (1), we obtain, for the development of log. $(1+x)$,

$$\log.\ (1+x)=\frac{\mathrm{A}}{1}.x-\frac{\mathrm{A}}{2}x^2+\frac{\mathrm{A}}{3}x^3-\frac{\mathrm{A}}{4}x^4+\ldots$$

$$=\mathrm{A}\left(\frac{x}{1}-\frac{x^2}{2}+\frac{x^3}{3}-\frac{x^4}{4}+\frac{x^5}{5}-\frac{x^6}{6}+\ldots\right).$$

The number A is called the *modulus* of the system of logarithms employed. Lord Napier, the illustrious inventor of logarithms, assumed the modulus equal to unity. If, then, we designate Naperian logarithms by log.′, we shall have

$$\log.\ (1+x)=\frac{x}{1}-\frac{x^2}{2}+\frac{x^3}{3}-\frac{x^4}{4}+\frac{x^5}{5}-\frac{x^6}{6}+,\ \&c.\quad (4).$$

By giving to x in succession all possible values, we may obtain from this equation the logarithms of all numbers.

If we make $x=0$, we shall have log.′ $1=0$.

Make $x=1$, and we obtain

$$\log.'\ 2=1-\tfrac{1}{2}+\tfrac{1}{3}-\tfrac{1}{4}+\tfrac{1}{5}-,\ \&c.,$$

a series which converges so slowly that it would be necessary to employ a very large number of terms to obtain the accuracy desirable. The series may be rendered more converging in the following manner:

In equation (4), substitute $-x$ for x, and it becomes

$$\log.'\ (1-x)=-\frac{x}{1}-\frac{x}{2}-\frac{x^3}{3}-\frac{x^4}{4}-,\ \&c.\quad (5).$$

Subtracting equation (5) from equation (4), and observing that $\log.'(1+x) - \log.'(1-x) = \log.'\frac{1+x}{1-x}$, we obtain

$$\log.'\left(\frac{1+x}{1-x}\right)=2\left(\frac{x}{1}+\frac{x^3}{3}+\frac{x^5}{5}+\frac{x^7}{7}+\frac{x^9}{9}+\ldots\right).$$

Put $x=\frac{1}{2z+1}$, and the preceding series becomes, by substitution,

$$\log.'\left(\frac{z+1}{z}\right)=\log.'(z+1)-\log.'z=2\left(\frac{1}{2z+1}+\frac{1}{3(2z+1)^3}+\frac{1}{5(2z+1)^5}+\ldots\right).$$

(348.) The last series may be employed for computing the logarithm of any number, when the logarithm of the preceding number is known. Making successively $z=1, 2, 4, 6$, &c., we obtain the following

NAPERIAN, OR HYPERBOLIC LOGARITHMS.

$\log.'\ 2=2(\frac{1}{3}+\frac{1}{3\cdot 3^3}+\frac{1}{5\cdot 3^5}+\frac{1}{7\cdot 3^7}+\ldots)$	$=0.693147$
$\log.'\ 3=\log.'2+2(\frac{1}{5}+\frac{1}{3\cdot 5^3}+\frac{1}{5\cdot 5^5}+\frac{1}{7\cdot 5^7}+\ldots)$	$=1.098612$
$\log.'\ 4=2\log.'2$	$=1.386294$
$\log.'\ 5=\log.'4+2(\frac{1}{9}+\frac{1}{3\cdot 9^3}+\frac{1}{5\cdot 9^5}+\frac{1}{7\cdot 9^7}+\ldots)$	$=1.609438$
$\log.'\ 6=\log.'3+\log.'2$	$=1.791759$
$\log.'\ 7=\log.'6+2(\frac{1}{13}+\frac{1}{3\cdot 13^3}+\frac{1}{5\cdot 13^5}+\frac{1}{7\cdot 13^7}+\ldots)$	$=1.945910$
$\log.'\ 8=3\log.'2$	$=2.079442$
$\log.'\ 9=2\log.'3$	$=2.197225$
$\log.'\ 10=\log.'5+\log.'2$	$=2.302585$
&c., &c.,	&c.

(349.) The Naperian logarithms being computed, it is easy to form any other system. We have found

$$\log.(1+x)=A\left(\frac{x}{1}-\frac{x^2}{2}+\frac{x^3}{3}-\frac{x^4}{4}+\frac{x^5}{5}-\frac{x^6}{6}+\ldots\right).$$

Distinguishing the Naperian logarithms by an accent, we have

$$\log.'(1+x)=A'\left(\frac{x}{1}-\frac{x^2}{2}+\frac{x^3}{3}-\frac{x^4}{4}+\frac{x^5}{5}-\frac{x^6}{6}\ldots\right).$$

Hence

$$\log.\ (1+x) : \log.'\ (1+x) :: A : A'.$$

Therefore, *the logarithms of the same number in different systems are to each other as the moduli.*

In Napier's system, the modulus $=1$. Hence

$$\log.\ (1+x) = A . \log.'\ (1+x).$$

That is, *the common logarithm of a number is equal to its Naperian logarithm multiplied by the modulus of the common system.*

If, then, we knew the modulus of the common system, we could easily convert the preceding Naperian logarithms into common logarithms. Now, from the equation

$$\log.\ (1+x) = A . \log'.\ (1+x), \text{ we obtair}$$

$$A = \frac{\log.\ (1+x)}{\log.'\ (1+x)}.$$

Suppose $x=9$, then $A = \dfrac{\log.\ 10}{\log.'\ 10}$.

But log. 10=1. Hence

$$A = \frac{1}{\log.'\ 10} = \frac{1}{2.302585} = 0.434294,$$

which is the modulus of the common system.

(350.) We can now compute the

COMMON, OR BRIGGS' LOGARITHMS.

log. 2=0.693147×0.434294	=0.301030
log. 3=1.098612×0.434294	=0.477121
log. 4=2 log. 2	=0.602060
log. 5= log. 10− log. 2=1− log. 2	=0.698970
log. 6= log. 3+ log. 2	=0.778151
log. 7=1.945910×0.434294	=0.845098
log. 8=3 log. 2	=0.903090
log. 9=2 log. 3.	=0.954243
log. 10=	=1.000000
&c., &c.,	&c.

We thus obtain the following Table of Common Logarithms:

No.	Logarithm.	No.	Logarithm.	No.	Logarithm.	No.	Logarithm.
1	0.000000	36	1.556303	71	1.851258	106	2.025306
2	0.301030	37	1.568202	72	1.857332	107	2.029384
3	0.477121	38	1.579784	73	1.863323	108	2.033424
4	0.602060	39	1.591065	74	1.869232	109	2.037426
5	0.698970	40	1.602060	75	1.875061	110	2.041393
6	0.778151	41	1.612784	76	1.880814	111	2.045323
7	0.845098	42	1.623249	77	1.886491	112	2.049218
8	0.903090	43	1.633468	78	1.892095	113	2.053078
9	0.954243	44	1.643453	79	1.897627	114	2.056905
10	1.000000	45	1.653213	80	1.903090	115	2.060698
11	1.041393	46	1.662758	81	1.908485	116	2.064458
12	1.079181	47	1.672098	82	1.913814	117	2.068186
13	1.113943	48	1.681241	83	1.919078	118	2.071882
14	1.146128	49	1.690196	84	1.924279	119	2.075547
15	1.176091	50	1.698970	85	1.929419	120	2.079181
16	1.204120	51	1.707570	86	1.934498	121	2.082785
17	1.230449	52	1.716003	87	1.939519	122	2.086360
18	1.255273	53	1.724276	88	1.944483	123	2.089905
19	1.278754	54	1.732394	89	1.949390	124	2.093422
20	1.301030	55	1.740363	90	1.954243	125	2.096910
21	1.322219	56	1.748188	91	1.959041	126	2.100371
22	1.342423	57	1.755875	92	1.963788	127	2.103804
23	1.361728	58	1.763428	93	1.968483	128	2.107210
24	1.380211	59	1.770852	94	1.973128	129	2.110590
25	1.397940	60	1.778151	95	1.977724	130	2.113943
26	1.414973	61	1.785330	96	1.982271	131	2.117271
27	1.431364	62	1.792392	97	1.986772	132	2.120574
28	1.447158	63	1.799341	98	1.991226	133	2.123852
29	1.462398	64	1.806180	99	1.995635	134	2.127105
30	1.477121	65	1.812913	100	2.000000	135	2.130334
31	1.491362	66	1.819544	101	2.004321	136	2.133539
32	1.505150	67	1.826075	102	2.008600	137	2.136721
33	1.518514	68	1.832509	103	2.012837	138	2.139879
34	1.531479	69	1.838849	104	2.017033	139	2.143015
35	1.544068	70	1.845098	105	2.021189	140	2.146128

(351.) Let us now determine the *base* of Napier's system. Designating it by a, we shall have, *Art.* 349,

$$\log.' a : \log. a :: 1 : 0.434294.$$

But $\log.' a = 1$. Hence

$$\log. a = 0.434294.$$

That is, *the modulus of the common system is equal to the common logarithm of Napier's base.*

We wish, then, to find the number corresponding to the common logarithm 0.434294. By inspecting the preceding table, we see that this number must be a little less than 3. More accurately, it is

$$2.718282,$$

which is the base of Napier's system.

Any number, except unity, may be taken as the base of a system of logarithms, and hence there may be an infinite number of systems. Only two systems, however, are much used; those of Briggs and Napier.

The base of	Briggs' system is		10.
"	Napier's	"	2.718282.
The modulus of	Briggs'	"	0.434294.
"	Napier's	"	1.

Hence, in Briggs' system, all numbers are to be regarded as powers of 10.

Thus,

$$10^{0.301}=2,$$
$$10^{0.477}=3,$$
$$10^{0.602}=4,$$
$$10^{0.698}=5,$$
&c., &c.

In Napier's system, all numbers are to be regarded as powers of 2.718282.

Thus,

$$2.718^{0.693}=2,$$
$$2.718^{1.098}=3,$$
$$2.718^{1.386}=4,$$
$$2.718^{1.609}=5,$$
&c., &c.

Briggs' logarithms are employed in all the common operations of multiplication and division, and hence they are known by the name of *common* logarithms. Napier's logarithms are of great use in the application of the calculus to many analytical and physical problems. They are also called *hyperbolic* logarithms, having been originally derived from the hyperbola.

EXPONENTIAL EQUATIONS.

(352.) An *exponential quantity* is one which is raised to some unknown power, or which has an unknown quantity for an exponent; as,

$$a^x,\ a^{\frac{1}{x}},\ x^x,\ \text{or}\ x^{\frac{1}{x}},\ \&c.$$

An *exponential equation* is one which contains an exponential quantity; as,

$$a^x=b,\ x^x=c,\ \&c.$$

Such equations are most easily solved by means of logarithms. Thus, consider the equation

$$a^x=b.$$

Taking the logarithm of each member of the equation, we have

$$x \log. a=\log. b,$$

$$\text{or } x=\frac{\log. b}{\log. a}.$$

Ex. 1. What is the value of x in the equation $3^x=81$?

By the preceding formula, $x=\frac{\log. 81}{\log. 3}$.

Looking out the logarithms of 81 and 3 from the Table on page 318, we have

$$x=\frac{1.908485}{.477121}=4.$$

Therefore, $3^4=81.$

Ex. 2. What is the value of x in the equation $3^x=20$?

$$x=\frac{\log. 20}{\log. 3}=\frac{1.301030}{.477121}=2.727 \text{ nearly.}$$

Therefore, $3^{2.727}=20$ nearly.

Ex. 3. What is the value of x in the equation $5^x=12$?

Ex. 4. What is the value of x in the equation $\left(\frac{2}{3}\right)^x=\frac{3}{4}$?

(353.) The other equation, $x^x=c$, may be solved by *trial*, as in *Art.* 333. Thus, taking the logarithm of each member, we have

$$x \log. x = \log. c.$$

Find now, by trial, two numbers nearly equal to the value of x, and substitute them for x in the given equation. Then say,

As the difference of these results,
Is to the difference of the two assumed numbers,
So is the error of either result,
To the correction required in the corresponding assumed number.

Ex. 1. Given $x^x=100$ to find the value of x.

Here we have $x \log. x = \log. 100 = 2.$

Suppose $x=3,$

then $0.477121 \times 3 = 1.431363$, which is *too small.*

Suppose $x=4,$

then $0.602060 \times 4 = 2.408240$, which is *too great.*

Hence the value of x is between 3 and 4, but nearer to 4 Assume, then, 3.5 and 3.6 for the two numbers.

By the first supposition,		*By the second supposition,*	
$x=3.5$; $\log. x=$	.544068	$x=3.6$; $\log. x=$	.556303
Multiplied by	3.5	Multiplied by	3.6
$x.\log. x=$	1.904238	$x.\log. x=$	2.002689

Diff. of results	:	*Diff. assumed numbers*	: :	*Error of 2d result*	:	*Its correction.*
.098451	:	0.1	: :	.002689	:	.00273

Hence $x=3.6-.00273=3.59727$ nearly.

Therefore, $3.59727^{3.59727}=100$ nearly.

If we wish a more accurate result, the operation must be repeated with two new numbers; as, for example, 3.59727 and 3.59728.

Ex. 2. Given $x^x=6$, to find the value of x.

Ex. 3. Given $x^x=20x$, to find an approximate value of x.

COMPOUND INTEREST.

(354.) In calculating compound interest, the first subject of inquiry is, to what sum does a given principal amount, after a certain number of years, the interest being annually added to the principal? It is evident that $1.00, placed out at 5 per cent., becomes, at the end of a year, a principal of $1.05. But the amount at the end of each year must be proportioned to the principal at the beginning of the year. In order, then, to find the amount at the end of two years, we institute the proportion

$$1.00 : 1.05 :: 1.05 : (1.05)^2.$$

The sum 1.05^2 must now be considered as the principal, and hence, to find the amount at the end of three years, we say

$$1.00 : 1.05 :: (1.05)^2 : (1.05)^3.$$

And in the same manner we find that the amount of $1.00 for n years at compound interest is $(1.05)^n$.

If the rate of interest were six per cent., we should find the amount for n years to be $(1.06)^n$.

The amount of two dollars for a given time must obviously be double the amount of one dollar, and the amount of $1000 must be a thousand times the amount of one dollar.

Hence, if we put P to represent the principal,
r the rate per cent. considered as a decimal,
n the number of years,
A the amount of the given principal for n years, we shall have

$$A = P.(1+r)^n.$$

This equation contains four quantities, A, P, n, r; any three of which being given, the fourth may be found. The computations are most readily performed by means of logarithms. Taking the logarithms of both members of the preceding equation, and reducing, we find

1. $\log. A = n \times \log. (1+r) + \log. P.$
2. $\log. P = \log. A - n \times \log. (1+r).$
3. $\log. (1+r) = \dfrac{\log. A - \log. P}{n}.$
4. $n = \dfrac{\log. A - \log. P}{\log. (1+r)}.$

EXAMPLES.

Ex. 1. What is the amount of twenty dollars, at 6 per cent. compound interest, for 11 years?

In this example we employ formula (1).

Amount of $1.00 for 1 year $1.06,	log. =0.025306
Multiplying by 11,	11
	0.278366
Given principal $20.	log. =1.301030
Amount $38 nearly,	1.579396

This result is derived from the Table on page 318. By consulting a larger Table, we should find the amount $37.97.

Ex. 2. What principal at 5 per cent. interest will amount to $66 in 13 years?

Here we employ formula (2).

$1+r=1.05$,	log. =0.021189
Multiplying by n,	13
Subtract	0.275457
From log. A,	1.819544
P=$35 nearly,	1.544087.

Ex. 3. At what rate per cent. must $40 be put out at compound interest, that it may amount to $57 in 9 years?

Here we employ formula (3).

A=57,	log. =1.755875
P=40,	log. =1.602060
Dividing by n,	9)0.153815
$1+r=1.04$	=0.017091

Consequently, $r=.04$, or four per cent.

How could this result be obtained without the use of logarithms?

Ex. 4. In what time will $50 amount to $90 at 5 per cent.

Here we employ formula (4).

A=90,	log. =1.954243
P=50,	log. =1.698970
$1+r=1.05$, whose logarithm is	0.021189)0.255273.

Dividing one logarithm by the other, we obtain 12, *Ans.*

Ex. 5. What is the amount of $52 at 3 per cent. compound interest for 15 years?

Ans. $81

Ex. 6. What principal at 6 per cent. compound interest wil amount to $101 in 4 years?

Ans. $80.

Ex. 7. At what rate will $10 amount to $16 in 16 years?

Ans. Three per cent

Ex. 8. What will $300 amount to in 10 years at compound interest semi-annually, the yearly rate being 6 per cent.?

Ex. 9. In what time will a sum of money double at 6 per cent. compound interest?

Ans. 11.89 years.

Ex. 10. In what time will a sum of money triple itself at 4 per cent. compound interest?

Ans. 28.01 years.

(355.) The natural increase of population in a country may be computed in the same way as compound interest. Knowing the population at two different dates, we compute the rate of increase by formula (3), and from this we may compute the population at any future time on the supposition of a uniform rate of increase.

EXAMPLES.

Ex. 1. The number of the inhabitants of the United States in 1790 was 3,900,000, and in 1840, 17,000,000. What was the average increase for every ten years?

Ans. 34 per cent.

Ex. 2. Suppose the rate of increase to remain the same for the next ten years, what would be the number of inhabitants in 1850?

Ans. 22,800,000.

Ex. 3. At the same rate, in what time would the number in 1840 be doubled?

Ans. 23.54 years.

Ex. 4. At the same rate, what was the population in 1780?

Ans. 2,900,000.

Ex. 5. At the same rate, in what time would the number in 1840 be tripled?

Ans. 37.31 years

MISCELLANEOUS EXAMPLES.

(356.) *Ex.* 1. Given $\left.\begin{array}{l}\frac{1}{x}+\frac{1}{y}=a,\\ \frac{1}{x}+\frac{1}{z}=b,\\ \frac{1}{y}+\frac{1}{z}=c,\end{array}\right\}$ to find the values of x, y, and z.

Ans. $x=\frac{2}{a+b-c}$, $y=\frac{2}{a-b+c}$, $z=\frac{2}{b+c-a}$.

Ex. 2. Given $\left.\begin{array}{l}x+y+z+t+u=25,\\ x+y+z+u+w=26,\\ x+y+z+t+w=27,\\ x+y+u+t+w=28,\\ x+z+u+t+w=29,\\ y+z+u+t+w=30,\end{array}\right\}$ to find the values of x, y, z, t, u, and w.

Ans. $x=3$, $y=4$, $z=5$, $u=6$, $t=7$, $w=8$.

Ex. 3. Given $\left.\begin{array}{l}x(x+y+z)=27,\\ y(x+y+z)=18,\\ z(x+y+z)=36,\end{array}\right\}$ to find the values of x, y, and z.

Ans. $x=3$, $y=2$, $z=4$.

Ex. 4. Given $\left.\begin{array}{l}xy=z,\\ yz=v,\\ xv=a,\\ yv=bx,\end{array}\right\}$ to find the values of x, y, z, and v.

Ans. $x=\frac{\sqrt{a}}{\sqrt[3]{b}}$, $y=\sqrt[3]{b}$, $z=\sqrt{a}$, $v=\sqrt{a}\sqrt[3]{b}$.

Ex. 5. Given $\left.\begin{array}{l}xyz=105,\\ xyv=135,\\ xzv=189,\\ yzv=315,\end{array}\right\}$ to find the values of x, y, z, and v.

Ans. $x=3$, $y=5$, $z=7$, $v=9$.

Ex. 6. Given $\left.\begin{aligned} x^2+\frac{x^4}{y^2}+y^2&=84, \\ x+\frac{x^2}{y}+y&=14, \end{aligned}\right\}$ to find the values of x and y.

Ans. $x=4$, $y=2$ or 8.

Ex. 7. Given $\dfrac{a-\sqrt{a^2-x^2}}{a+\sqrt{a^2-x^2}}=b$, to find the values of x.

Ans. $x=\pm\dfrac{2a\sqrt{b}}{1+b}$.

Ex. 8. Given $\dfrac{\sqrt{x}+\sqrt{x-a}}{\sqrt{x}-\sqrt{x-a}}=\dfrac{ab^2}{x-a}$, to find the values of x.

Ans. $x=\dfrac{a(1\pm b)^2}{1\pm 2b}$.

Ex. 9. Given $\left.\begin{aligned} \sqrt{y}-\sqrt{a-x}&=\sqrt{y-x}, \\ 2\sqrt{y-x}+2\sqrt{a-x}&=5\sqrt{a-x}, \end{aligned}\right\}$ to find the values of x and y.

Ans. $x=\frac{4}{5}a$, $y=\frac{5}{4}a$.

Ex. 10. Given $\dfrac{\sqrt{a+x}}{\sqrt{x}}+\dfrac{\sqrt{a-x}}{\sqrt{x}}=\sqrt{\dfrac{x}{b}}$, to find the values of x.

Ans. $x=\pm 2\sqrt{ab-b^2}$.

Ex. 11. Given $\left.\begin{aligned} x^3+xy^2&=ay, \\ x^2y+y^3&=bx, \end{aligned}\right\}$ to find the values of x and y.

Ans. $x=\sqrt{\dfrac{\sqrt{a^3b}}{a+b}}$, $y=\sqrt{\dfrac{\sqrt{ab^3}}{a+b}}$

Ex. 12. Given $\dfrac{a^2+ax+x^2}{a+x}+\dfrac{a^2-ax+x^2}{a-x}=\dfrac{ab}{3a-4b+x}$, to find the values of x.

Ans. $x=-3a$, or $3a-\dfrac{2a^2}{b}$.

Ex. 13. Given $\dfrac{a-x}{b+x}-\dfrac{b-x}{a+x}=\dfrac{a+b}{a-b}$, to find the values of x

Ans. $x=\dfrac{\pm\sqrt{5}(a-b)-(a+b)}{2}$

Ex. 14. Given $\frac{\sqrt{1+x}}{1+\sqrt{1+x}}=\frac{\sqrt{1-x}}{1-\sqrt{1-x}}$, to find the values of x.

Ans. $x=\pm\frac{1}{2}\sqrt{3}$.

Ex. 15. Given $\frac{a+x+\sqrt{2ax+x^2}}{a+x}=b$, to find the values of x.

Ans. $x=\frac{\pm a(1\mp\sqrt{2b-b^2})}{\sqrt{2b-b^2}}$.

Ex. 16. Given $\left.\begin{array}{l}\sqrt{5\sqrt{x}+5\sqrt{y}}+\sqrt{x}+\sqrt{y}=10,\\ \sqrt{x^5}+\sqrt{y^5}=275,\end{array}\right\}$ to find the values of x and y.

Ans. $x=9$, $y=4$.

Ex. 17. Given $\left.\begin{array}{l}x-y+\sqrt{\frac{x-y}{x+y}}=\frac{12}{x+y},\\ x^2+y^2=41,\end{array}\right\}$ to find the values of x and y.

Ans. $x=5$, $y=4$.

Ex. 18. Given $\left.\begin{array}{l}(x+y)^3+x+y=30,\\ x-y=1,\end{array}\right\}$ to find the values of x and y.

Ans $x=2$, $y=1$.

Ex. 19. Given $x^4-4x^3+7x^2-6x=18$, to find the values of x by a quadratic equation.

Ans. $x=3$, or -1.

Ex. 20. Given $\left.\begin{array}{l}(x+y)(xy+1)=18xy,\\ (x^2+y^2)(x^2y^2+1)=208x^2y^2,\end{array}\right\}$ to find the values of x and y.

Ans. $x=2\pm\sqrt{3}$, $y=7\pm4\sqrt{3}$.

Ex. 21. Given $\left.\begin{array}{l}(x^2+y^2)xy=13090,\\ x+y=18,\end{array}\right\}$ to find the values of x and y.

Ans. $x=7$, or 11,
$y=11$, or 7.

Ex. 22. Given $\left.\begin{array}{l}5(x^2+y^2)+4xy=356,\\ x^2+y^2+x+y=62,\end{array}\right\}$ to find the values of x and y.

Ans. $x=4$, $y=6$.

Ex. 23. Given $\left.\begin{array}{l}(x^2+y^2)xy=300,\\ x^4+y^4=337,\end{array}\right\}$ to find the values of x and y.

Ans. $x=4$, $y=3$

Ex. 24. Given $(x^2+y^2)(x^3+y^3)=455,\ x+y=5,$ to find the values of x and y.

Ans. $x=3,\ y=2.$

Ex. 25. Given $\frac{x^2+xy+y^2}{x+y}=14,$ and $\frac{x^2-xy+y^2}{x-y}=18,$ to find the values of x and y.

Ans. $x=12,\ y=6.$

Ex. 26. Given $(x^2-xy+y^2)(x^2+y^2)=91,\ (x^2-xy+y^2)(x^2+xy+y^2)=133,$ to find the values of x and y.

Ans. $x=2$, or -3; $y=3$, or -2.

Ex. 27. Given $(x+y)xy=30,\ (x^2+y^2)x^2y^2=468,$ to find the values of x and y.

Ans. $x=2,\ y=3.$

Ex. 28. The sum of two numbers is a, and the sum of their reciprocals is b. Required the numbers.

Ans. $\frac{a}{2}\pm\sqrt{\frac{a^2}{4}-\frac{a}{b}}.$

Ex. 29. In the composition of a certain quantity of gunpowder, the nitre was ten pounds more than two thirds of the whole; the sulphur was four and a half pounds less than one sixth of the whole; and the charcoal was two pounds less than one seventh of the nitre. How many pounds of gunpowder were there?

Ans. 69 pounds.

Ex. 30. Find three numbers such that if six be subtracted from the first and second, the remainders will be in the ratio of 2 : 3; if thirty be added to the first and third, the sums will be in the ratio of 3 : 4; but if ten be subtracted from the second and third, the remainders will be as 4 : 5.

Ans. 30, 42, 50.

Ex. 31. Divide the number 165 into five such parts that the first increased by one, the second increased by two, the third diminished by three, the fourth multiplied by 4, and the fifth divided by five, may all be equal.

Ans 19, 18, 23, 5, and 100

Ex. 32. A criminal having escaped from prison, traveled ten hours before his escape was known. He was then pursued, so as to be gained upon three miles an hour. After his pursuers had traveled eight hours, they met an express going at the same rate as themselves, who met the criminal two hours and twenty-four minutes before. In what time from the commencement of the pursuit will they overtake him?

Ans. 20 hours.

Ex. 33. A and B engage to reap a field of wheat in twelve days. The times in which they could severally reap an acre are as 2 : 3. After some days, finding themselves unable to finish it in the stipulated time, they call in C to help them, whose rate of working was such that, if he had wrought with them from the beginning, it would have been finished in nine days. Also, the times in which he could have reaped the field with A alone, and with B alone, are in the ratio of 7 : 8. When was C called in?

Ans. After six days.

Ex. 34. A laborer is engaged for n days, on condition that he receives p pence for every day he works, and pays q pence for every day he is idle. At the end of the time he receives a pence. How many days did he work, and how many was he idle?

Ans. He worked $\frac{nq+a}{p+q}$, and was idle $\frac{np-a}{p+q}$ days.

Ex. 35. The fore wheel of a carriage makes three revolutions more than the hind wheel in going sixty yards; but, if the circumference of each wheel be increased one yard, it will make only two revolutions more than the hind wheel in the same space. Required the circumference of each.

Ans. 4 and 5 yards.

Ex. 36. There is a wagon with a mechanical contrivance by which the difference of the number of revolutions of the wheels on a journey is noted. The circumference of the fore wheel is a feet, and of the hind wheel b feet. What is the distance gone over when the fore wheel has made n revolutions more than the hind wheel?

Ans. $\frac{abn}{b-a}$ feet.

Ex. 37. A merchant has two casks, each containing a certain quantity of wine. In order to have an equal quantity in each, he pours out of the first cask into the second as much as the second contained at first; then he pours from the second into the first as much as was left in the first; and then again, from the first into the second as much as was left in the second, when there are found to be a gallons in each cask. How many gallons did each cask contain at first?

Ans. $\frac{11a}{8}$ and $\frac{5a}{8}$.

Ex. 38. A and B engage to reap a field for \$24; and as A alone could reap it in nine days, they promise to complete it in five days. They found, however, that they were obliged to call in C to assist them for the last two days, in consequence of which B received one dollar less than he otherwise would have done. In what time could B or C alone reap the field?

Ans. B in 15, and C in 18 days.

Ex. 39. A cistern can be filled by four pipes; by the first in a hours, by the second in b hours, by the third in c hours, and by the fourth in d hours. In what time will the cistern be filled when the four pipes are opened at once?

Ans. $\frac{abcd}{abc+abd+acd+bcd}$.

Ex. 40. The sum of the cubes of two numbers is 35, and the sum of their ninth powers is 20195. Required the numbers.

Ans. 2 and 3.

Ex. 41. A number consisting of three digits, which are in Arithmetical Progression, being divided by the sum of its digits, gives a quotient 26; and if 198 be added to it, the digits will be inverted. Required the number.

Ans. 234.

Ex. 42. There are three numbers in Geometrical Progression, the difference of whose differences is six, and their sum is forty-two. Required the numbers.

Ans. 6, 12, and 24.

Ex. 43. There are three numbers in harmonical proportion;

the sum of the first and third is 18, and the product of the three numbers is 576. Required the numbers.

Ans. 12, 8, and 6.

Ex. 44. There are three numbers in harmonical proportion, the difference of whose differences is 2, and four times the product of the first and third is 960. Required the numbers.

Ans. 20, 15, 12.

Ex. 45. There are two numbers whose product is 300; and the difference of their cubes is thirty-seven times the cube of their difference. What are the numbers?

Ans. 20 and 15.

Ex. 46. There are three numbers in geometrical progression, the greatest of which exceeds the least by 24; and the difference of the squares of the greatest and the least is to the sum of the squares of all the three numbers as 5 : 7. What are the numbers?

Ans. 8, 16, and 32.

Ex. 47. A merchant had $26,000, which he divided into two parts, and placed them at interest in such a manner that the incomes from them were equal. If he had put out the first portion at the same rate as the second, he would have drawn for this part $720 interest; and if he had placed the second out at the same rate as the first, he would have drawn for it $980 interest. What were the two rates of interest?

Ans. 6 per cent. for the larger sum, and 7 for the smaller.

Ex. 48. A grocer has a cask containing 20 gallons of brandy, from which he draws off a certain quantity into another cask of equal size, and, having filled the last with water, the first cask was filled with the mixture. It now appears that if $6\frac{2}{3}$ gallons of the mixture are drawn off from the first into the second cask, there will be equal quantities of brandy in each. Required the quantity of brandy first drawn off.

Ans. 10 gallons.

Ex. 49. A miner bought two cubical masses of ore for $820. Each of them cost as many dollars per cubic foot as there were feet in a side of the other; and the base of the greater contained a square yard more than the base of the less. What was the price of each?

Ans. 500 and 320 dollars.

Ex. 50. A and B traveled on the same road, and at the same rate, from Cumberland to Baltimore. At the 50th milestone from Baltimore, A overtook a drove of geese, which were proceeding at the rate of three miles in two hours; and two hours afterward met a wagon, which was moving at the rate of nine miles in four hours. B overtook the same drove of geese at the 45th milestone, and met the same wagon 40 minutes before he came to the 31st milestone. Where was B when A reached Baltimore?

Ans. 25 miles from Baltimore.

Ex. 51. A gentleman bought a rectangular lot of land at the rate of ten dollars for every foot in the perimeter. If the same quantity had been in a square form, and he had bought it at the same rate, it would have cost him $330 less; but if he had bought a square piece of the same perimeter, he would have had $12\frac{1}{4}$ rods more. What were the dimensions of the lot?

Ans. 9 by 16 rods.

Ex. 52. A and B put out at interest sums amounting to $2400. A's rate of interest was one per cent. more than B's; his yearly interest was five sixths of B's; and at the end of ten years his principal and simple interest amounted to five sevenths of B's. What sum was put at interest by each, and at what rate?

Ans. A $960, at 5 per cent.
B $1440, at 4 "

Ex. 53. Two merchants sold the same kind of cloth. The second sold three yards more of it than the first, and together they received $35. The first said to the second, I should have received $24 for your cloth; the other replied, I should have received $$12\frac{1}{2}$ for yours. How many yards did each of them sell?

Ans. The first merchant, 5 or 15 yards
The second " 8 or 18 "

Ex. 54. A person bought a quantity of cloth of two sorts for $63. For every yard of the best piece he gave as many dollars as he had yards in all; and for every yard of the poorer, as many dollars as there were yards of the better piece more than of the poorer. Also, the whole cost of the best piece was six times that of the poorer. How many yards had he of each?

Ans. 6 yards of the better, and 3 of the poorer.

Ex. 55. A and B, 165 miles distant from each other, set out with a design to meet. A travels 1 mile the first day, 2 the second, 3 the third, and so on. B travels 20 miles the first day, 18 the second, 16 the third, and so on. In how many days will they meet?

Ans. 10 or 33 days.

Ex. 56. There are three numbers in Geometrical Progression whose continued product is 216, and the sum of their cubes is 1971. Required the numbers.

Ans. 3, 6, and 12.

Ex. 57. There are four numbers in Geometrical Progression whose sum is 350; and the difference between the extremes is to the difference of the means as 37 : 12. What are the numbers?

Ans. 54, 72, 96, 128.

Ex. 58. A commences a piece of work alone, and labors for two thirds of the time that B would have required to perform the entire work. B then completes the job. Had both labored together, it would have been completed two days sooner; and A would have performed only half what he left for B. Required the time in which they would have performed the work separately.

Ans. A in 6 days, and B in 3 days.

Ex. 59. A ship, with a crew of 175 men, set sail with a supply of water sufficient to last to the end of the voyage; but in 30 days the scurvy made its appearance, and carried off three men every day; and at the same time a storm arose which protracted the voyage three weeks. They were, however, just enabled to arrive in port without any diminution in each man's daily allowance of water. Required the time of the passage, and the number of men alive when the vessel reached the harbor.

Ans. The voyage lasted 79 days, and the number of men alive was 28.

Ex. 60. The number of deaths in a besieged garrison

amounted to 6 daily; and, allowing for this diminution, their stock of provisions was sufficient to last 8 days. But on the evening of the sixth day 100 men were killed in a sally, and afterward the mortality increased to 10 daily. Supposing the stock of provisions unconsumed at the end of the sixth day to support 6 men for 61 days, it is required to find how long it would support the garrison, and the number of men alive when the provisions were exhausted.

Ans. The provisions last 6 days, and 26 men survive.

Note.—Hints for the solution of Ex. 20, p. 327.

Expanding equation first, and dividing by xy, we obtain

$$x+\frac{1}{x}+y+\frac{1}{y}=18.$$

In a similar manner, we obtain from equation second

$$x^2+\frac{1}{x^2}+y^2+\frac{1}{y^2}=208.$$

Put
$$x+\frac{1}{x}=v,$$
$$y+\frac{1}{y}=w,$$

and we have
$$v+w=18,$$
$$v^2+w^2=212.$$

Whence
$$v=14 \text{ or } 4,$$
$$w=4 \text{ or } 14;$$

and the values of x and y are easily found.

THE END.

A text-book like this of Prof. Loomis was much needed, and the desideratum is so well supplied, that I think it can not fail to commend itself at once to the favorable regard of those who are looking for the best work for college classes. I consider it decidedly the best book for college instruction that I am acquainted with on the subject, and it has been adopted as a text-book in our college by unanimous consent of the faculty. Prof. Loomis has been very happy in simplifying the more difficult parts of the subject, especially on the theory of equations and on logarithms. —JAMES NOONEY, A.M., *Professor of Mathematics and Nat. Philosophy in Western Reserve College.*

Prof. Loomis' work is well calculated to impart a clear and correct knowledge of the principles of Algebra. The rules are concise, yet sufficiently comprehensive, containing in few words all that is necessary, and *nothing more;* the absence of which quality mars many a scientific treatise. The collection of problems is peculiarly rich, adapted to impress the most important principles upon the youthful mind, and the student is led gradually and intelligently into the more interesting and higher departments of the science.—JOHN BROCKLESBY, A.M., *Professor of Mathematics and Natural Philosophy in Trinity College.*

I am much pleased with Prof. Loomis' Algebra. I think he has accomplished very happily the object he had in view, and has prepared a work remarkably well adapted for the use of college students.—E. S. SNELL, A.M., *Professor of Mathematics and Natural Philosophy in Amherst College*

I am much pleased with Prof. Loomis's Algebra. The arrangement of the subjects is, I think, an admirable one. The best proof I can give of the estimation in which I hold it is, that I am now hearing recitations in it the second time.—JOHN TATLOCK, A.M., *Professor of Mathematics in Williams College.*

I have examined Prof. Loomis' Algebra with great attention, and am so well pleased with its arrangement and execution throughout, that I have decided to adopt it as a text-book in this institution.—THOMAS E. SUDLER, A.M., *Professor of Mathematics in Dickinson College.*

Prof. Loomis has here aimed at exhibiting the first principles of Algebra in a form which, while level with the capacity of ordinary students and the present state of the science, is fitted to elicit that degree of effort which educational purposes require. Throughout the work, whenever it can be done with advantage, the practice is followed of generalizing particular examples, or of extending a question proposed relative to a *particular* quantity, to the *class* of quantities to which it belongs, a practice of obvious utility, as accustoming the student to pass from the particular to the general, and as fitted to impress a main distinction between the literal and numeral calculus. The general doctrine of Equations is expounded with clearness, and, we may add, with independence. The author has developed this subject in an order of his own. Theorems which find a place in other treatises are omitted, and what sometimes appears in a generic form, or in that of a corollary, becomes specific, or assumes the place of a primary proposition. We venture to say that there will be but one opinion respecting the general character of the exposition. —*American Journal of Science and Arts.*

I regard Prof. Loomis' Algebra as altogether worthy of the high reputation its author deservedly enjoys. It possesses those qualities which are chiefly requisite in a college text-book. Its statements are clear and definite; the more important principles are made so prominent as to arrest the pupil's attention; and it conducts the pupil by a sure and easy path to those habits of *generalization* which the teacher of Algebra has so much difficulty in imparting to his pupils.—JULIAN M. STURTEVANT, LL.D., *President of Illinois College.*

The arrangement of Prof. Loomis' Algebra is good; the doctrine of Equations is clearly presented, and the principle of generalization is ably developed in a manner well calculated to improve the youthful mind.—W. P. ALRICH, *Professor of Mathematics in Washington College, Pa.*

Prof. Loomis' Algebra is admirably got up. It is clear and simple in arrangement, and just the work for the class of learners for whom the author prepared it. The introduction of Horner's admirable method for finding incommensurable roots, and the section on Logarithms, render it superior to any text-book on this subject with which I am acquainted.—Pres. COLLINS, *Emory and Henry College, Virginia.*

We feel bound to express our conviction that this is a decidedly better text-book, especially for those not already far advanced in the study, than any other we have seen. It is carefully and lucidly arranged, and admirably enunciated and explained.—*Teachers' Journal.*

The present work is the fruit of long experience in teaching and diligent investigation of the science. The author has sought to avoid unnecessary prolixity on the one hand, and undue brevity on the other, and with the observance of this happy medium he has embodied all the latest improvements.—*Methodist Quarterly Review.*

LOOMIS' GEOMETRY AND CONIC SECTIONS.

Third Edition. 8vo, p. 226, Sheep, $1 00.

Every page of this book bears marks of careful preparation. Only those propositions are selected which are most important in themselves, or which are indispensable in the demonstration of others. The propositions are all enunciated with studied precision and brevity. The demonstrations are complete without being encumbered with verbiage; and, unlike many works we could mention, the diagrams are good representations of the objects intended. We believe this book will take its place among the best elementary works which our country has produced.—*American Review.*

Prof. Loomis' Geometry is characterized by the same neatness and elegance which were exhibited in his Algebra. While the logical form of argumentation peculiar to Playfair's Euclid is preserved, more completeness and symmetry is secured by additions in solid and spherical geometry, and by a different arrangement of the propositions. It will be a favorite with those who admire the chaste forms of argumentation of the old school; and it is a question whether these are not the best for the purposes of mental discipline.—*Northern Christian Advocate.*

As a text-book for instruction, this work possesses advantages superior, in some respects, to any other work on the subject in our language. The arrangement is good, and the selection of propositions so judiciously made, as to comprise what is most valuable for the purposes of the student, both for intellectual culture and for a knowledge of geometry. Prof. Loomis has introduced some valuable improvements, especially that of computing the area of a circle in a very simple and easy manner, and that of shading the diagrams in solid geometry, which will greatly aid the student in forming his conceptions of solid angles and the positions of planes.—*Ohio Observer.*

The enunciations in Prof. Loomis' Geometry are concise and clear, and the processes neither too brief nor too diffuse. The part treating of solid geometry is undoubtedly superior, in clearness and arrangement, to any other elementary treatise among us.—*New York Evangelist.*

Prof. Loomis has admirably harmonized the logical system of the Greek geometer with the more rapid processes of modern mathematicians.—*New York Observer.*

Having been requested, by a resolution of our Board of Trustees, to report such a course of mathematical studies as I may deem best suited to our circumstances, I have selected Loomis' Geometry and Conic Sections as a part of the course.—Matthew J. Williams, *Professor of Mathematics in South Carolina College.*

Prof. Loomis has made Legendre's Geometry far more *Euclidian*, and therefore more valuable. Some of his demonstrations are decided improvements on those of Legendre.—Professor C. Dewey, *Rochester University.*

This book is far in advance of Playfair's Euclid. It can not fail to come into general use.—*Albany Atlas.*

I consider Loomis' Geometry and Trigonometry the best works that I have ever seen on any branch of elementary mathematics.—James B. Dodd, A.M., *Professor of Mathematics, Transylvania University.*

Having used Loomis' Elements of Geometry for two years, carefully examined it, and compared it with Euclid and Legendre, I have found it preferable to either. Teachers will find the work an excellent text-book, suited to give a clear view of the beautiful science of which it treats.—Alonzo Gray, A.M., *Professor of Mathematics, Brooklyn Female Academy.*

This work is admirably drawn up, and merits universal adoption in all schools where these branches of science are taught.—*Courier and Enquirer.*

These books are terse in style, clear in method, easy of comprehension, and perfectly free from that useless verbiage with which it is too much the fashion to load school-books under pretense of explanation.—*Scott's Weekly Paper, Canada.*

Prof. Loomis is doing a valuable service to the cause of mathematical science by the course of text-books he is preparing. His writings in other departments of science are characterized by a remarkable clearness in the manner in which he exhibits truth, and his treatises on Algebra and Geometry bear evident marks of having emanated from the same mind.—James H. Coffin, A.M., *Professor of Mathematics in Lafayette College.*

Prof. Loomis has made many improvements in Legendre's Geometry, retaining all the merits of that author without the defects. I shall adopt his work as a text-book in this college.—Thomas E. Sudler, A.M., *Professor of Mathematics in Dickinson College.*

LOOMIS' TRIGONOMETRY AND TABLES.

8vo, p. 320, Sheep, $1 50.

This work contains an exposition of the nature and properties of Logarithms; the principles of Plane Trigonometry; the Mensuration of Surfaces and Solids; the principles of Land Surveying, with a full description of the instruments employed; the Elements of Navigation, and of Spherical Trigonometry. The Tables furnish the Logarithms of Numbers to 10,000, with the proportional parts for a fifth figure in the natural number; logarithmic Sines and Tangents for every ten Seconds of the Quadrant, with the proportional parts to single seconds; natural Sines and Tangents for every Minute of the Quadrant; a Traverse Table; a Table of Meridional Parts, &c.

The following are a few of the notices of this work which have been received by the publishers:

Loomis' Trigonometry is well adapted to give the student that distinct knowledge of the principles of the science so important in the further prosecution of the study of mathematics. The description and representation of the instruments used in surveying, leveling, &c., are sufficient to prepare the student to make a practical application of the principles he has learned. The Tables are just the thing for college students.—JOHN TATLOCK, A.M., *Professor of Mathematics in Williams College.*

Prof. Loomis has done up the work admirably. The brevity and clearness which characterize this excellent system of mathematical reasoning are the *ne plus ultra* for such a work. His Trigonometry will meet with the approval already accorded to his Algebra and Geometry.—Professor C DEWEY, *Rochester University.*

Loomis' Trigonometry is sufficiently extensive for collegiate purposes, and is every where clear and simple in its statements without being redundant. The learner will here find what he really needs without being distracted by what is superfluous or irrelevant.—A. CASWELL, D.D., *Professor of Mathematics and Natural Philosophy in Brown University.*

Loomis' Tables are vastly better than those in common use. The extension of the sines and tangents to ten seconds is a great improvement. The tables of natural sines are indispensable to a good understanding of Trigonometry; and the natural tangents are exceedingly convenient in analytical geometry.—I. WARD ANDREWS, A.M., *Professor of Mathematics and Natural Philosophy in Marietta College.*

Loomis' Trigonometry and Tables are a great acquisition to mathematical schools. I know of no work in which the principles of Trigonometry are so well condensed and so admirably adapted to the course of instruction in the mathematical schools of our country. I shall adopt it as a text-book for instruction in this college.—THOMAS E. SUDLER, A.M., *Professor of Mathematics in Dickinson College.*

Loomis' Trigonometry and Tables are both excellent works, and I shall recommend them at every opportunity which offers.—JAMES CURLEY, *Professor in Georgetown College.*

I am so much pleased with Prof. Loomis' Trigonometry that I intend to use it as a text-book in this college.—JOHN BROCKLESBY, A.M., *Professor of Mathematics in Trinity College.*

In this work the principles of Trigonometry and its applications are discussed with the same clearness that characterizes the previous volumes. The portion appropriated to Mensuration, Surveying, &c., will especially commend itself to teachers, by the judgment exhibited in the extent to which they are carried, and the practically useful character of the matter introduced. What I have particularly admired in this, as well as the previous volumes, is the constant recognition of the difficulties, present and prospective, which are likely to embarrass the learner, and the skill and tact with which they are removed. The Logarithmic Tables will be found unsurpassed in practical convenience by any others of the same extent.—AUGUSTUS W. SMITH, LL.D., *Professor of Mathematics and Astronomy in the Wesleyan University.*

Prof. Loomis' text-books in Mathematics are models of neatness, precision, and practical adaptation to the wants of students.—*Methodist Quarterly Review.*

LOOMIS' ELEMENTS OF ANALYTICAL GEOMETRY, AND OF THE DIFFERENTIAL AND INTEGRAL CALCULUS.

8vo, p. 278, Sheep, $1 50.

THIS treatise constitutes the fourth volume of a course of Mathematics designed for colleges and high schools. The first part treats of the application of Algebra to Geometry, the construction of equations, the properties of a straight line, a circle, parabola, ellipse, and hyperbola; the classification of algebraic curves, and the more important transcendental curves. The second part treats of the differentiation of algebraic functions, of Maclaurin's and Taylor's theorems, of maxima and minima, transcendental functions, theory of curves, and evolutes The third part exhibits the method of obtaining the integrals of a great variety of differentials, and their application to the rectification and quadrature of curves, and the cubature of solids. All the principles are illustrated by an extensive collection of examples. The work was prepared to meet the wants of the mass of college students of average abilities.

The following are a few of the notices of this work which have been received by the publishers:

I have examined Loomis' Analytical Geometry and Calculus with great satisfaction, and shall make it an indispensable part of our scientific course.—JAMES B. DODD, A.M., *Professor of Mathematics in Transylvania University.*

Loomis' Analytical Geometry and Calculus is the best work on that subject for a college course and mathematical schools. It contains all the important principles and doctrines of the calculus, simplified and illustrated by well-selected problems.—THOMAS E. SUDLER, A.M., *Professor of Mathematics in Dickinson College.*

Loomis' Calculus is better adapted to the capacities of young men than any book heretofore published on this subject.—A. P. HOOKE, *Professor of Mathematics in Bethany College.*

Loomis' Analytical Geometry and Calculus is prepared with the same careful regard for the actual wants of the mathematical student, the same vigilant eye to the difficulties which are most likely to beset him in his progress, that molded the previous volumes. The author has made a valuable contribution to the means of education, and indirectly to the cause of science—not by the invention of new methods of research, but by opening the door to a larger number; making, indeed, this apartment of the temple of knowledge accessible to all who wish to possess themselves of one of the most efficient instruments of investigation, or who are moved only by a desire to gratify a reasonable curiosity. In no part of a mathematical study was simplification, and clear, palpable illustration more urgently called for than in this, and in no American work has this object been more satisfactorily accomplished.—*Middletown (Connecticut) Sentinel.*

This is a work which can not fail to be popular with the mass of students. It will be found every thing the author intended it to be, and can not be too warmly commended.—*Albany Spectator.*

The Analytical Geometry is treated, amply enough for elementary instruction, in the short compass of 112 pages, so that nothing need be omitted, and the student can master his text-book as a whole. The Calculus is treated in like manner in 167 pages, and the opening chapter makes the nature of the art as clear as it can possibly be made. We recommend this work, without reserve or limitation, as the best text-book on the subject we have yet seen.—*Methodist Quarterly Review.*

The processes are clear and regular, and the examples well arranged to illustrate the principles.—*Christian Register.*

I am well pleased with Loomis' Analytical Geometry and Calculus, as it brings the subjects within the powers of the majority of our students, a thing certainly that very few authors on the Calculus try to do.—JAMES CURLEY, *Professor of Mathematics in Georgetown College.*

Loomis' Calculus is to the purpose for colleges, and we shall use it.—CHESTER DEWEY, *Professor in Rochester University.*

LOOMIS' RECENT PROGRESS OF ASTRONOMY,

ESPECIALLY IN THE UNITED STATES.

12mo, p. 258, Muslin, $1 00.

THIS volume is designed to exhibit in a popular form the most important astronomical discoveries of the past ten years. It treats particularly of the discovery of the planet Neptune, of the new asteroids, of the new satellite of Saturn, of the great comet of 1843, Biela's comet, Miss Mitchell's comet, &c.; of the parallax of fixed stars, motion of the stars, resolution of nebulæ, &c.; the history of American observatories, determination of longitude by the electric telegraph, manufacture of telescopes in the United States, &c.

The following are a few of the notices of this work which have been received.

Loomis' "Recent Progress of Astronomy" has afforded me great interest, for it is admirably done. As a work to be read by a multitude of our intelligent people who are not adepts in astronomy, it has no competitor. It supplies a desideratum that was strongly felt, and must gratify numbers who are interested in the progress of astronomy in our own country.—CHESTER DEWEY, *Professor in Rochester University.*

In the "Recent Progress of Astronomy" we have the model of a class of works which we deem of great importance to the popular diffusion of scientific knowledge. Without sacrificing any thing of mathematical exactness, conceding no claim of the most rigid investigation for the sake of immediate effect, and employing no trickery of method or style to attract attention, it is strictly a popular treatise, presenting the results of protracted and extensive research in language of transparent simplicity, and placing the difficult topics which it discusses in a light which makes them comprehensible by the generality of intelligent readers. The author writes from that fullness of knowledge which enables him to make a compact and lucid statement of the point under consideration. Professor Loomis is eminently happy in seizing on the most essential points, and unfolding them with a clearness and precision which make his work no less readable than it is instructive.—*New York Tribune.*

The "Recent Progress of Astronomy" is a work which very fully and exactly meets the wants of educated men. It gives full and clear views of various subjects connected with astronomy, about which many who have received an elevated education entertain only obscure and unsatisfactory notions. The chapter relating to the progress of astronomy in the United States is peculiarly interesting. The public are under great obligations to Professor Loomis for giving in so attractive a form the results of the labors of men of science.—*New York Observer.*

The "Progress of Astronomy" is written with great clearness, and can not fail to be read by all students of this sublime science.—*Evening Journal.*

Professor Loomis could hardly have given to the people a more acceptable and useful work. This book is designed for the public, and may be read with profit by those who have never studied the simple elements of the science.—*Watchman and Reflector.*

This is a beautifully printed and illustrated volume, detailing and explaining in clear and intelligible terms the recent discoveries and advances in this noble science. The volume is well adapted to the general reader, as the use of technicalities has been avoided as far as was consistent with accuracy.—*Saroni's Musical Times.*

A very attractive book, and one which can not be read without profit.—*Springfield Republican.*

The narratives which this work contains of the many important discoveries in the field of astronomical labor are of the most interesting character. They are written in a pleasing style, and are in a great degree free from the dry details of complex calculations which the uninitiated would fail to understand.—*Boston Daily Journal.*

The design of this book is most happily carried out. The unlearned reader can have little difficulty in following its luminous expositions; and the interest of the subject is so great that few who begin the perusal of the book will fail to finish it.—*Methodist Quarterly Review.*

The design of this work will at once commend it to public favor. It narrates in simple and untechnical terms the most important discoveries which have been made within the last ten years, and especially notices the recent efforts for improvement and advancement in the noble science in this country.—*New York Evangelist.*

LOOMIS' ELEMENTS OF ALGEBRA.

DESIGNED FOR BEGINNERS.

12mo, Sheep, p. 260, price 62½ cents.

This volume is intended for the use of students who have just completed the study of Arithmetic. It is believed that it will be found sufficiently clear and simple to be adapted to the wants of a large class of students in our common schools. It explains the method of solving equations of the first degree, with one, two, or more unknown quantities; the principles of involution and evolution; the solution of equations of the second degree; the principles of ratio and proportion; with arithmetical and geometrical progression. Every principle is illustrated by a copious collection of examples, and a variety of miscellaneous problems will be found at the close of the book.

The following are a few of the notices of this work which have been received by the publishers:

I have used Loomis' Elements of Algebra in my school for the last six months, and have found it fitted in a high degree to give the pupil a clear and sufficiently comprehensive knowledge of the elements of the science. I believe teachers of Academies and High Schools will find it all that they can desire as a text-book on this branch of Mathematics.—Prof. ALONZO GRAY, *Brooklyn Heights Seminary.*

I am so much pleased with Loomis' Elements of Algebra that I have introduced it as a text-book in the Institution under my care.—Rev. GORHAM D. ABBOTT, *Spingler Institute.*

Loomis' Elements of Algebra is worthy of adoption in our Academies, and will be found to be an excellent text-book. The definitions and rules are expressed in simple and accurate language; the collection of examples subjoined to each rule is sufficiently copious; and as a book for beginners it is admirably adapted to make the learner thoroughly acquainted with the first principles of this important branch of science.—D. MACAULEY, *Principal of the Polytechnic School, New Orleans.*

I am very highly pleased with Loomis' series of Mathematical text-books. The gentleman who has charge of the mathematical department in this Institution says that I can hardly speak too favorably of its merits. We shall adopt it forthwith.—CHARLES F. DEEMS, *President of Greensboro Female College, N. C.*

Loomis' Algebras form an excellent progressive course for the young student. The "Elements" could be put with advantage into the hands of every child who has mastered the principles of Arithmetic, and is admirably adapted for the use of common schools. The explanations and simplifications of the author are extremely lucid and comprehensive.—*N. Y. Observer.*

The favorable impressions made on my mind by a perusal of Loomis' Mathematics have been fully realized by my experience in the use of them in the recitation-room.—JOSHUA O. COLBURN, *Principal of Lancaster Academy, Penn.*

I have carefully examined Loomis' Elements of Algebra, and cheerfully recommend it on account of its superior arrangement, and clear and full explanations.—SOLOMON JENNER, *Principal of N. Y. Commercial School.*

Loomis' Elements of Algebra is prepared with the care and judgment that characterize all the elementary works published by the same author.—*Methodist Quarterly Review.*

Professor Loomis's Mathematical Series.

THE RECENT PROGRESS OF ASTRONOMY; especially in the United States. 12mo, Muslin.

A TREATISE ON ALGEBRA. 8vo, Sheep. $1 00.

ELEMENTS OF GEOMETRY AND CONIC SECTIONS. 8vo, Sheep. $1 00.

ELEMENTS OF PLANE AND SPHERICAL TRIGONOMETRY, with their Applications to Mensuration, Surveying, and Navigation. 8vo, Sheep. $1 00.

TABLES OF LOGARITHMS OF NUMBERS, and of Sines and Tangents for every ten Seconds of the Quadrant. 8vo, Sheep. $1 00.

☞ The TRIGONOMETRY and TABLES, bound together, may be had for $1 50.

Professor M'Clintock's School Classics.

FIRST BOOK IN LATIN. Containing Grammar, Exercises, and Vocabularies. 12mo, Sheep. 75 cents.

A SECOND BOOK IN LATIN. Containing a complete Latin Syntax, with copious Exercises for Imitation and Repetition, &c. 12mo. [In press.]

AN INTRODUCTION TO LATIN STYLE. Principally from the German of Grysar. 12mo. [In press.]

A FIRST BOOK IN GREEK. Containing full Vocabularies, Lessons on the Forms of Words, &c. 12mo, Sheep. 75 cents.

A SECOND BOOK IN GREEK. Containing the Syntax and Prosody, with Reading Lessons, and forming a sufficient Greek Reader. 12mo, Sheep. 75 cents.

Professor Hackley's Mathematical Series.

A TREATISE ON ALGEBRA. Containing the latest Improvements. 8vo, Sheep. $1 50.

A SCHOOL ALGEBRA. Containing the latest Improvements. 8vo, Muslin. 75 cents.

AN ELEMENTARY COURSE OF GEOMETRY, for the Use of Schools and Colleges. 12mo, Sheep. 75 cents.

Professor Parker's English School Series.

AIDS TO ENGLISH COMPOSITION. Prepared for the Student of all Grades. 12mo, Sheep, 90 cents; Muslin, 80 cents.

GEOGRAPHICAL QUESTIONS. Adapted for the Use of Morse's or most other Maps. 12mo, Muslin. 25 cents.

OUTLINES OF GENERAL HISTORY, in the Form of Question and Answer. 12mo, Sheep. $1 00.

Professor Renwick's Philosophical Works.

FIRST PRINCIPLES OF CHEMISTRY. With Questions. Engravings. 18mo, half Sheep. 75 cents.

THE SCIENCE OF MECHANICS APPLIED TO PRACTICAL PURPOSES. Engravings. 18mo, half Roan. 75 cents.

FIRST PRINCIPLES OF NATURAL PHILOSOPHY. With Questions. Engravings. 18mo, half Roan. 75 cents.

Professor Boyd's Academical Works.

ECLECTIC MORAL PHILOSOPHY. Prepared for Literary Institutions and General Use. 12mo, Sheep, 87½ cents; Muslin, 75 cents.

ELEMENTS OF RHETORIC AND LITERARY CRITICISM. 12mo, half Bound, 50 cents.

Dr. Abercrombie's Philosophical Works.

TREATISE ON THE INTELLECTUAL POWERS. With Questions. 18mo, half Roan, 50 cents; Muslin, 45 cents.

PHILOSOPHY OF THE MORAL FEELINGS. With Questions. 18mo, half Roan, 50 cents; Muslin, 45 cents

Professor Draper's Philosophical Works.

A TEXT-BOOK OF CHEMISTRY. With nearly 300 Illustrations. 12mo, Sheep. 75 cts.

A TEXT-BOOK OF NATURAL PHILOSOPHY. With nearly 400 Illustrations. 12mo, Sheep. 75 cents.

CHEMICAL ORGANIZATION OF PLANTS. Engravings. 4to. $2 50.

Professor Upham's Philosophical Series.

ELEMENTS OF MENTAL PHILOSOPHY. 2 vols. 12mo, Sheep. $2 50.

An Abridgment of the Above, by the Author. 12mo, Sheep. $1 25.

PHILOSOPHICAL AND PRACTICAL TREATISE ON THE WILL. 12mo, Sheep extra. $1 25.

OUTLINES OF IMPERFECT AND DISORDERED MENTAL ACTION. 18mo, Muslin. 45 cents.

Dictionaries and Works of Reference.

A DICTIONARY OF THE ENGLISH LANGUAGE. By Noah Webster, LL.D A new Edition, revised and enlarged, by C. A. Goodrich, Professor in Yale College. 8vo, Sheep. $3 50.

LIDDELL AND SCOTT'S NEW GREEK AND ENGLISH LEXICON. Based on the German Work of Passow. With Additions, &c., by H. Drisler, under the Supervision of Prof. Anthon. Royal 8vo, Sheep extra. $5 00.

An Abridgment of the Above, by the Authors, for the Use of Schools, with the Addition of a Second Part, viz., English and Greek. [In press.]

RIDDLE AND ARNOLD'S ENGLISH-LATIN LEXICON. Founded on the German-Latin Dictionary of Dr. C. E. Georges. First American Edition, carefully revised, and containing a Dictionary of Proper Names, by Charles Anthon, LL.D. Royal 8vo, Sheep extra. $3 00.

A UNIVERSAL GAZETTEER, Geographical, Statistical, and Historical. By J. R. M'Culloch. Edited by D. Haskel, A.M. 2 vols. 8vo, Sheep, $6 50; Muslin, $6 00.

ENCYCLOPEDIA OF SCIENCE, LITERATURE, AND ART. Edited by W. T. Brande, assisted by J. Cauvin. Engravings. 8vo, Sheep. $4 00.

THE NORTH AMERICAN ATLAS. Containing 36 Folio Maps in Colors, forming a complete Atlas of this Continent. By S. E. Morse. Half Roan. $2 75.

ENGLISH SYNONYMS EXPLAINED. With copious Illustrations and Explanations. By G. Crabbe, M.A. 8vo, Sheep extra. $2 00.

A CLASSICAL DICTIONARY. Containing an Account of the principal Proper Names mentioned in Ancient Authors. By C. Anthon. 8vo, Sheep. $4 00.

DICTIONARY OF GREEK AND ROMAN ANTIQUITIES. Edited by W. Smith. Enlarged, &c., by C. Anthon. Engravings. 8vo, Sheep. $4 00.

An Abridgment of the Above, by the Authors. 12mo, half Sheep. 90 cents.

THE ENGLISHMAN'S GREEK CONCORDANCE OF THE NEW TESTAMENT. 8vo, Sheep extra, $5 00; Muslin, $4 50.

POTTER'S HAND-BOOK FOR READERS AND STUDENTS. 18mo, Muslin. 45 cents

Miscellaneous Works.

ROBINSON'S GREEK-ENGLISH LEXICON OF THE NEW TESTAMENT. 8vo, Sheep. $4 75.

BUTTMANN'S GREEK GRAMMAR. Edited by Dr. Robinson. 8vo, Sheep

GRAY'S ELEMENTS OF NATURAL PHILOSOPHY. 12mo, Sheep. 75 cents.

FOWLER'S TREATISE ON THE ENGLISH LANGUAGE, in its Elements and Forms. Including a full Grammar. 8vo, Sheep.

SMITH'S ELEMENTARY TREATISE ON MECHANICS. 8vo, Sheep extra, $1 75; Muslin, $1 50.

WHATELY'S ELEMENTS OF LOGIC. 18mo, Muslin. 38 cents.

WHATELY'S ELEMENTS OF RHETORIC. Complete. 18mo, Muslin. 38 cents.

HOBART'S ANALYSIS OF BUTLER'S ANALOGY OF RELIGION. Also, Crauford's Questions for Examination. Revised and adapted to the Use of Schools. By C. E. West. 18mo, Muslin. 40 cents.

MARKHAM'S HISTORY OF FRANCE, from the Conquest of Gaul by Julius Cæsar to the Reign of Louis Philippe. Prepared for the Use of Schools, by J. Abbott. 12mo, half Sheep. $1 25.

SCHMITZ'S HISTORY OF ROME. With Questions by J. ROBSON. 12mo, Muslin. 75 cts

THE NEW TESTAMENT IN GREEK, with English Notes, Critical, Philological, and Exegetical Indexes, &c., by Rev. J. A. SPENCER. 12mo, Sheep, $1 40; Muslin, $1 25.

MILL'S LOGIC, RATIOCINATIVE AND INDUCTIVE. 8vo, Muslin. $2 00.

THE BOTANY OF THE UNITED STATES, North of Virginia. With a Sketch of the Rudiments of Botany, and a Glossary of Terms. By LEWIS C. BECK. 12mo, Muslin. $1 25.

THE JUVENILE SPEAKER. Comprising Exercises in Declamation. By F. T. RUSSELL. 12mo, half Bound, 70 cents; Muslin, 60 cents.

HARPER'S NEW YORK CLASS-BOOK. A Reading book for Schools. By W. RUSSELL. 12mo, half Sheep. $1 25.

MORSE'S NEW PICTORIAL SCHOOL GEOGRAPHY. 150 Engravings, and about 50 Maps in Colors. 4to. 50 cents.

A FIRST BOOK IN SPANISH. Adapted to every Class of Learners. By J. SALKELD. 12mo, Sheep, $1 25; Muslin, $1 00.

A COMPENDIUM OF ROMAN AND GRECIAN ANTIQUITIES, including a Sketch of Ancient Mythology, Maps, &c. By J. SALKELD. 12mo, Muslin. 37½ cents.

THE NORTH AMERICAN ACCOUNTANT. By P. DUFF. Royal 8vo, half Bound. 75 cents.

ELEMENTS OF ALGEBRA, embracing also the Theory and Application of Logarithms, &c. By D. W. CLARK. 8vo, Sheep. $1 00.

POLITICAL ECONOMY. With a Summary for the Use of Students. By A. POTTER 18mo, half Roan, 50 cents; Muslin, 45 cents.

BURKE'S ESSAY ON THE SUBLIME AND BEAUTIFUL. Edited by A. MILLS. 12mo, Muslin. 75 cents.

ALISON ON THE NATURE AND PRINCIPLES OF TASTE. Edited by A. MILLS. 12mo, Muslin. 75 cents.

ELEMENTS OF MORALITY, INCLUDING POLITY. By W. WHEWELL. 2 vols. 12mo, Muslin. $1 00.

A SYSTEM OF INTELLECTUAL PHILOSOPHY. By A. MAHAN. 12mo, Muslin. 90 cents.

THE PHILOSOPHY OF RHETORIC. By G. CAMPBELL. 12mo, Muslin. $1 25.

LETTERS ON ASTRONOMY. Addressed to a Lady. By D. OLMSTED. With numerous Engravings. 12mo, Muslin. 75 cents.

BOURCHARLAT'S ELEMENTARY TREATISE ON MECHANICS. Translated and edited by E. H. COURTENAY. Plates. 8vo, Sheep. $2 25.

PLATO AGAINST THE ATHEISTS; or, the Tenth Book of the Dialogue on Laws, &c. By TAYLER LEWIS. 12mo, Muslin. $1 50.

THE CAPTIVES. A Comedy of PLAUTUS. With English Notes, for the Use of Students, by J. PROUDFIT. 18mo, Paper. 37½ cents.

EPITOME OF THE HISTORY OF PHILOSOPHY. Translated from the French, with Additions, by C. S. HENRY, D D. 2 vols. 18mo, Muslin. 90 cents.

ROBERTSON'S HISTORY OF THE DISCOVERY OF AMERICA. With Questions, by J. FROST, A.M. Engravings. 8vo, Sheep extra. $1 75.

ROBERTSON'S HISTORY OF THE EMPEROR CHARLES V. With Questions, by J. FROST, A.M. Engravings. 8vo, Sheep extra. $1 75.

TREATISE ON DOMESTIC ECONOMY. For the Use of Young Ladies at Home and at School. By Miss C. E. BEECHER. Revised Edition. 12mo, Muslin, gilt. 75 cents.

PRINCIPLES OF PHYSIOLOGY. By A. COMBE, M.D. With Questions. Engravings. 18mo, half Sheep, 50 cents; Muslin, 45 cents.

ANIMAL MECHANISM AND PHYSIOLOGY. By J. H. GRISCOM, M.D. Engravings. 18mo, half Sheep, 50 cents; Muslin, 45 cents.

GLASS'S LIFE OF WASHINGTON, in Latin Prose. Edited by J. N. REYNOLDS. Portrait. 12mo, Muslin. $1 12½.

HEMPEL'S GRAMMAR OF THE GERMAN LANGUAGE. Arranged into a new System on the Principle of Induction. 2 vols. 12mo, half Bound. $1 75.

LEE'S ELEMENTS OF GEOLOGY. Containing a Description of the Geological Formations and Mineral Resources of the United States. Engravings. 18mo, half Sheep. 50 cents.

KANE'S ELEMENTS OF CHEMISTRY; including the most recent Discoveries and Applications of the Science to Medicine and Pharmacy, and to the Arts. Edited by JOHN W. DRAPER, M.D. With about 250 Wood-cuts. 8vo, Sheep, $2 25; Muslin, $2 00.

MAURY'S PRINCIPLES OF ELOQUENCE. With an Introduction, by the Rev. Dr. POTTER. 18mo, Muslin. 45 cents.

SCHMUCKER'S PSYCHOLOGY; or, Elements of a new System of Mental Philosophy, on the basis of Consciousness and Common Sense. Designed for Colleges and Academies. 12mo, Muslin. $1 00.

www.ingramcontent.com/pod-product-compliance
Lightning Source LLC
LaVergne TN
LVHW010157110826
845151LV00002B/541
* 9 7 8 1 4 2 5 5 3 8 1 9 4 *